Foreword

Let's talk about the world.

Specifically, let's talk about the *historical narrative*. By "historical narrative," I mean our general unified consensus of how the world works. That may seem like an oddly clumsy way of describing something, but when you think about it, it's really quite accurate, isn't it? We don't really **know** for sure exactly how the world works. We **think** we know – we assemble a mental picture that we are pretty confident about – but we could be totally wrong, couldn't we? After all, our ancestors also thought they had a good mental picture of how the world worked, but **they** turned out to be totally wrong, didn't they? Good old chief Montezuma of the Aztecs thought that it was only his human sacrifices that kept the sun from falling, but he turned out to be totally incorrect about that, wasn't he? If some benevolent deity were to somehow transport him to the present day, he'd sure have to revise his mental model. As the primitive chief gazed in slack-jawed wonder at the marvels of civilization – slowly taking the whole thing in - you might even say that he was *updating* his backwards worldview to the general unified consensus of how the world works. The new *historical narrative*, in other words.

But consensus changes, doesn't it? I mean, we can never really be **sure** that we've fully understood the true objective reality of how the world works, or even how ethics work. It's all basically subjective if you really stop to think about it. According to Silicon Valley visionaries like Elon Musk, we could all potentially be virtual characters in a simulation. Or mental stimuli being sent into a brain in a jar. Or the dreams of some ancient gods. Elon Musk waxes poetic about a lot of stuff, so I hope you'll forgive me if I haven't characterized his views too accurately.

My point is that there **is** no "end of history," no "final end zone" where we no longer are capable of making earth-shattering discoveries that alter our perception of the *historical narrative*. If we are ever going to advance the future of science and our species, we need to be prepared to accept the fact that for all we know, a lot of our scientific and cultural beliefs could turn out to be just total nonsense – as outdated and out of touch with reality as chief Montezuma's view of the world. Are we prepared to do that, to evolve beyond our current understanding of the world? Or are we going to stop evolving and die? Because those are the only two choices, folks. The only **real** law – the only part of our *historical narrative* that has continually stood the test of time – is the law of the jungle. Evolve continuously or eventually die out, once our species runs into something it can't handle.

I for one would prefer to evolve. And if you are reading this book of your own free will, then I'm guessing you have chosen to evolve also. So read on.

Chapter 1
Peeking through the Great Lie
How to identify and terminate personal self-delusion

HOW LIFE EVOLVED

Let's talk about how life evolved. This is not news to any of you readers, I'm sure – after all, we've all taken high-school biology, and in my opinion the *historical narrative* is pretty accurate in this regard. But one thing we haven't examined are the implications from a psychological and developmental perspective, so I'd like to spend some time reviewing the beginnings of life again in a generally condensed way, before we then drill down to the practical applications of this knowledge.

Naturally, we are not the only ones to ask these questions. A new Silicon Valley startup named Cambrian Genomics recently set its eyes on the admirable goal of creating life. But when you think of it from the abstract perspective of a physicist, creating life is easy. All you have to do is leave a **lot** of hydrogen in one place for a really long time, adding heat to gradually accelerate the chemical reactions. Eventually the hydrogen will chemically react with other naturally occurring impurities present in the mixture, starting a chemical process that leads to hydrocarbons and other similar structures. Once you get enough hydrocarbons doing their non-sentient dance, eventually they will create life. In fact, from an extremely long-term perspective, some might even say that it's harder NOT to create life than it is to create it. The nature of the universe itself (a starry void full of giant flaming balls of hydrogen) means that the creation of life was inevitable, given enough time. Now consider the implications of that. When the first life was created, there was no arbitrary line where a "divine spark" was created out of increasingly complex chemical processes. Can **you** put your finger on the exact point where a bunch of random chemical processes "become alive?" And ask yourself this – if we can't put our finger on the exact point in time where *life* begins, then how can we ever put our finger on the exact moment in time where *consciousness* begins? How do we even know that we have consciousness at all? That we are not simply the continuation of a long series of chemical processes that only mimics the illusion of consciousness?

I'll get back to that in a later chapter.

Eventually, these complex arrangements of hydrocarbons form cells, and at this point we start to recognize the beginning of that vague boundary which our scientists generally define as life. There is still no perfect consensus on this, and there are some edge cases that many scientists are still unclear on, such as viruses. Viruses are considered edge cases because they do not display many of the "self-directed" characteristics of life, being unable to reproduce on their own as "living" organisms do. In fact, they are nothing more than chemical patterns that life absorbs and gets changed by, replicating additional chemical patterns (of the same type as the original) as a side effect of the change. This *grey area* between life and death becomes very important in later chapters, but for now it suffices to say that the most precise definition of life would be "any complex combination of self-reproducing patterns larger than the viral level." That makes sense because it neatly characterizes the difference between life and death. When you are alive, your chemical patterns (ie, the hydrocarbons that form the building blocks of your cells) *self-reproduce*. When you die, your chemical patterns stop *self-reproducing*. You want to identify the only real difference between life and death, folks? Well, that's the best answer that modern science has given us thus far. In life we are made up of chemical patterns that self-reproduce, in death we are made up of chemical patterns that do not. Fundamentally, that is the best guess that modern science has ever been able to make.

At least, under the current *historical narrative*.

Anyway, once we get into cells, things get a bit more predictable, because unlike the randomness of death, life is orderly. Life has *goals*. Life wants more of itself, and Life wants to avoid Death. And this trait gets stronger and stronger with each passing generation, because the Life that isn't driven hard enough by those two psychological imperatives (its chemical patterns, if you will) generally doesn't survive long enough to pass on its chemical patterns so that the next set of Life can grow up harder, better, faster, stronger. So you can generally assume that as Darwinism (our only real law, if you'll recall) takes its toll on each subsequent generation of Life, future generations will become more and more adept at evolving traits which benefit those two objectives, or they will die.

Anyway, that's the abstract, high-level summary. Now let's go into the practical applications of this knowledge. How is it *useful* to be able to predict the future direction that life will evolve in? Well, for starters, if you see two species occupying the same ecological niche, you can generally look at the environment and make an educated guess about what optimal traits will be useful to flourish in that environment. From there it's not too hard to extrapolate which of the two species will survive and which will gradually die off.

Wait a second, most people don't **have** several million years to hang around making large bets on which species will survive, so that's not exactly *useful* by itself. Still, it's an interesting tidbit of information that might be a *useful* tool to understanding other concepts, so let's just store that idea in our mental toolbox for now.

THE NATURE OF IDEAS

There are two *historical narratives* when it comes to *ideas*. Currently the dominant *historical narrative* is that we human beings think about various things and events, and these thoughts give birth to *ideas*, which we then transmit to each other through the medium of language. The central hypothesis that I am putting forwards in this book is that the people who believe this *historical narrative* are all incorrect.

There is a less-vocal but much more scientifically-oriented historical narrative which suggests that ideas are more like a murmuration of starlings. In this alternate *historical narrative*, the only reason we evolved intelligence was as a tool to help us survive. I mean, let's be really honest here – in the Darwinian sense, our minds are really our only advantage against the many other predators in the animal kingdom, aren't they? Apes have more muscle fibers, leopards have those claws, alligators have their teeth… but what do humans have? We evolved to favor one attribute – one evolved characteristic – to the exclusion of all others. That trait is our intelligence. Human beings became the dominant predators on the entire planet because we kill things with our minds, through the intermediary device of technology. Under this model of the world, our minds are a parliament of ideas –

aggression, love, creativity, dominance, etc – all vying for possession of our bodies at any one time. Over time, the ideas that prove most useful to a specific environment become better at possessing us, and this causes a feedback loop that results in certain behavioral traits achieving greater prominence. But that's a complicated explanation. The reductive way to describe it would be that certain ideas are better suited to some environments than others, and as these ideas evolve over the millennia the ones which are most optimized towards a specific environment get ingrained into their hosts. On the national level, we would refer to this as *culture*.

Obviously, I tend to favor the second *historical narrative* which is happily scientific, while also depicted very accurately in the adorable children's movie Inside Out. But in Disney's narrative, the ideas stay inside the head of the person that they are piloting. What if that was not necessarily the case? What if ideas could be transmitted from person to person, like a virus? In fact, what if that happens **every single day of our lives** on the subconscious level - and we are simply unaware of this process consciously. Viral ideas... hmmm, where have I heard the concept of "an idea going viral" before?

Oh, that's right, I believe it's a field that came under a certain degree of scrutiny during the 2016 U.S. presidential election. My esteemed colleagues, this field is called memetics – the replication of ideas. Memetics is one of the many fields of study in an area of scholarship that I term "The Dark Arts of Rationality." There are many ways to define the high-minded intellectual field of "Rationality," but the definition I personally favor is "the study of ideas." Unfortunately, science is a dull academic pursuit and tends to over-intellectualize problems, reducing them to abstract arguments that can never be fully resolved, tested, or measured. It tends to draw a certain starchy type of personality; you know what I mean? Other personality types like to dive in and get our hands dirty tinkering around under the hood of this new *historical narrative*. We like to take the experimental hypotheses out for a test drive just to see how the wheels spin. That's why I refer to this field of scholarship as the **Dark** Arts of Rationality: because if you want to use these techniques, you might accidentally end up getting your hands a bit dirty with engine grease, so to speak. Pure rationalists are scientists who develop *theories*; whereas we "Dark Rationalists" are simply rationalists who take a more pragmatic engineering-oriented approach and convert *theories* into *tools*.

So let's try a thought experiment: what if we don't actually have ideas? What if the **ideas** have **us**?

HOW IDEAS EVOLVE

Do you remember way back in this chapter where we defined life as "any complex combination of self-reproducing patterns larger than the viral level?" Well, doesn't that basically describe the structure of any type of *knowledge*? Information is a **pattern** encoded into data, regardless of whether said data is disseminated via gossip, imagery, art, or song. Imagine that complicated packets of information – ie, *knowledge*, also known as *ideas* – are the DNA equivalent of the chemical patterns that constitute our physical lives. An egregore, in other words.

Under such an unusual (but compelling) *historical narrative*, what would memes be analogous to? In my view, memes would be similar to viruses because they embody that *grey area* we mentioned earlier which is not quite life, not quite death. After all, memes don't replicate on their own – they only spread when a medium of information arises that can transmit them.

Of course, this is assuming that we view egregores as being the equivalent of simple life forms, analogous to a unicellular organism. When an idea is considered from the perspective of it being a microscopic level organism, then by comparison memes would be smaller and less self-directed than the idea itself, so one could think of the meme as a virus. But it is entirely possible to take an entirely different perspective and view an idea as a highly sophisticated and evolved lifeform. Under such an advanced schema, we could best understand memes as the natural evolved defenses of an egregore – its claws and teeth, as it were. We could even take a more macroscopic view and conceive of an egregore as a sort of "hive-mind," such as that murmuration of starlings that I mentioned earlier. Under such a model, each individual who willingly believes the idea - who allows it to occupy space in their minds – would be the equivalent of one individual starling in the murmuration.

The reason it is *useful* to model ideas as egregores is twofold. First, if ideas are alive, then we ought to be able to predict their behavior and evolution; because Life is orderly and has *goals*, and the magic of Darwinism ensures that those *goals* are the same for all types of life. Second, if ideas are alive, then we may be able to interact with them and to a limited extent attempt to direct their behavior, exactly the same way as we currently interact with all other life on the planet.

For example, politics on an abstract sense can be considered a clash of ideas because different political platforms have different ideologies. So one can generally make an informed estimate about which ideologies are going to survive based on the memetic patterns embedded in each ideology's culture. Some might liken this to ancient animal fights at the Roman Coliseum which would often pit different kinds of exotic animals against each other – a bear versus a lion, elephant versus rhinoceros, and such. By examining each animal's natural defenses – claws, teeth, muscle distribution, padded hide, etc – a spectator at these games could make a very reasonable guess about which animal species was most likely to win. We can make very similar bets with clashes between competing ideologies simply by examining each ideology closely to see what natural memes it has evolved as a defense mechanism over time.

This may sound good in theory, but how do we recognize this principle in practice?

Our first example of what egregores doing battle (a literal battle of ideas) might look like comes from World War 2. According to our current *historical narrative*, Nazism was a terrible ideology rooted in evil and the forces of good won out, as they always do. Throughout this book, you will notice that a theme about "the forces of good inevitably winning" recurs persistently in our current *historical narrative*. This is because of the substantial influence that Hollywood mythology has in today's culture: culture is a large factor in what influences our *historical narrative*.

In order to advance our scientific knowledge as swiftly as possible, let's assume that we are all completely non-racist people who have spent our last two hours doing the obligatory performative condemnation of Nazism so that we can move on to categorizing and studying it not as an **ideology**, but rather as a **living creature**. I want to run you through an anatomy lesson on this egregore so that we can identify its strengths and weaknesses. For the purposes of categorizing Nazism memetically, let us consider any characteristic that either enhances the ideology's ability to replicate in people's minds or improves the well-being of its host culture (since that incentivizes adoption of the ideology) to be a strength. We will consider any characteristic that either reduces the ideology's ability to replicate or reduces the well-being of its host culture (since that disincentives adoption) to be a weakness. In other words, we are assuming egregores follow the same incentives as all life – namely, the urge to replicate their patterns as efficiently as possible while avoiding obsolescence. Life wants more of itself, and Life wants to avoid Death.

<u>Nazism (Anatomical study)</u>

<u>Strengths</u>	<u>Weaknesses</u>
• Generates cultural unity by channeling anger towards external threats.	• Lack of genetic and ideological diversity due to focus on ethnic and cultural purity.
• Enhanced social trust results in stronger social safety nets.	• Non-whites have no incentive to buy into an ideology that places them at the bottom of the ideology's status hierarchy.
• Expansionistic and colonialist agenda results in rapid replication of ideology through military force.	

While it may seem tasteless or heretical in our current culture of fashionable outrage to suggest that Nazism might have had some strengths as well as noticeably glaring flaws, it is very important to classify the memetic traits associated with an ideology as precisely and scientifically as possible. This is because accurately identifying all the traits and characteristics associated with the ideology that you are researching allows you to ascertain its environmental niche in the ecology of egregores. For example, somebody from ancient Rome might have never seen or heard of a tiger until visiting a gladiatorial match for the first time, but by looking at its teeth and claws they could probably figure out very easily that it wasn't an herbivore. Similarly, we can look at the "teeth and claws" of an egregore (its **memes**, in other words) and establish what the egregore's "reproductive" (or **replicative**) life strategy might be.

For example, by looking at the Strengths column in the chart above, we can pretty easily establish that Nazism is not exactly what one would term a "friendly" ideology. The memes associated with it are all associated with the aggressive imposition of the ideology through militaristic means. In other words, this egregore is a predator.

Similarly, by examining the Weaknesses column, we can determine the shortcomings that might hold this ideology back, or serve as potential exploits for the ideology's enemies. For example, if Nazism had somehow managed to survive WW2, a lack of genetic diversity as well as their cultural obsession with Aryanism would most likely have resulted in major health problems among that ideology's host population due to inbreeding with a strong selection bias towards recessive genes. (There is a certain comedic irony in the concept of a eugenics-obsessed ideology that accidentally wipes itself out through its own eugenics - the tragedy of the Hapsberg dynasty writ large across an entire nation.) Similarly, a lack of exposure to other ideas and belief systems limited Nazi science from reaching its full potential, because important scientific discoveries can arise from scientists of **any** race or culture, and by artificially restricting their pool of effective scientists to those of non-Jewish Caucasian descent, Nazis ended up losing access to some of the most important scientific discoveries that ended up changing the course of the war, such as the atomic bomb. Consider that Albert Einstein was a German Jew who ended up fleeing Germany when he saw the tide of public opinion moving against the Jews, whom Hitler had unfairly branded as a privileged race that oppressed German society. If it were not for the anti-Semitic tribal memes embedded throughout Nazi ideology - which Einstein was smart enough to accurately recognize as a foreshadowing of Germany's future – Germany would have had the atomic bomb instead of the U.S., and the course of history would have been forever changed. In other words, if we consider Nazism to be an egregore – a living ideology - it should also be obvious that it was the egregore's evolutionary weaknesses which ended up destroying it. Germany lost the science race because they missed out on two of the brightest scientific minds of that era – both of whom happened to be Jews. One could say that a cancerous ideology that persecutes its own geniuses eventually destroys itself. Another way to phrase it might be that because there is no *incentive* for visionaries to serve an egregore which preys upon them, these architects of the future have a tendency to defect to the service of any rival egregore which protects and values them, an egregore which promises to reward their loyal service. A third way to describe it might be that if you build a shitty system, it will have shitty consequences that create a *reaction force* back upon you.

IDEAS AS POWER

If we operate under the *historical narrative* that egregores exist, and that a recent conflict between them was enough to shape the course of the entire planet… these things are starting to sound pretty *powerful* now, aren't they? Earlier in this chapter, I speculated on whether an egregore could best be modelled as a unicellular organism, a complex animal, or a murmuration group mind. If we perceive the ideology that we are tracking down to be an animal, then observing this animal using the memetic equivalent of claws to carve brand-new geopolitical boundaries into the globe seems like a pretty *sophisticated* kind of behavior, wouldn't you say? If we had previously thought that the egregore we were hunting for was the equivalent of a unicellular organism, seeing *sophisticated* behavior like this might lead us to *update* our assumptions.

So if we operate according to the *historical narrative* that living ideas exist, and that they can have world-shaping abilities, how can we harness that power? After all, a car that we can't drive isn't particularly *useful* to us. Two potential uses leap immediately to mind: predicting the future and shaping the future.

Predicting the future is relatively straightforward. If we can model group ideas - whether these ideas are things like political ideologies, corporate cultures, or municipal governance – relatively accurately as egregores, then being able to determine which egregore has more evolutionary advantages allows us to predict the rise and fall of political groups, corporations, and cities. This is *useful* because not only can we potentially make a fortune trading stocks and investing in real estate, we can even predict which political ideologies will eventually become dominant and associate ourselves with those movements early, so that we have more influence in the direction that they evolve.

Shaping the future is more complex, though there are people with a natural gift for it. Since memes are the teeth and claws of an egregore, then people with the skill to embed new memes into a culture can channel its evolution by implanting traits that make it stronger and more resilient. Another way to frame it is that by altering an egregore's natural defenses to enhance its Darwinian fitness, a person with a talent for memetics can tip the balance of conflict between competing ideologies.

IDEAS AS A PATTERN OR FREQUENCY

Of all the ways to model egregores, I think that the most accurate may be as a murmuration. The idea can best be visualized as a wave of sound flowing across a vast field of tuning forks, which represent our minds. For some people in that crowd, the idea will resonate. A stronger and more compelling idea will resonate at multiple frequencies, thus setting more "tuning forks" humming as crowds absorb the new idea. If enough people's minds are set to vibrating in alignment with the new idea, then the *historical narrative* changes. Remember that we always define the *historical narrative* as "our general unified consensus of how the world works." By extension, if the consensus changes, then the *historical narrative* changes as well.

The reason this knowledge is *useful* is because having a more accurate *historical narrative* is a source of great power. Science is part of any significant narrative shift, and it is easier to make scientific discoveries when operating within a more modern paradigm. This is important because thanks to the magic of Darwinism, civilizations and political movements that are operating in an outdated narrative (aka "extended failure mode") tend to be rapidly destroyed by their competition. Since there are a lot of evolutionary advantages to a culture which updates to a better narrative, and few (if any) disadvantages, it is in everybody's best interest to update to the most accurate *historical narrative* as frequently as possible. After all, societies lost in total self-delusion tend not to be particularly healthy, and this has a sad cost in human suffering that can be seen every day in the slums, orphanages, and hospitals of our world.

With that said, perhaps it would be useful to reexamine our existing narratives of the world, looking for some spots where our *historical narrative* might be faulty and might cause us unnecessary hardship.

THE HEROIC NARRATIVE

When comparing other historical cultures to our own, one thing that really stood out to me about the differences in our narratives is how we relate to heroes. In more ancient cultures, being a hero was not something that was given a significant moral weight. For example, in Greek myths, Hector was on one side, Achilles was on the other, and they ended up doing terrible things to each other. Yet both of them are portrayed as being heroic people. By contrast, the modern myths that come of Hollywood are much more binary, with a certain set of values that we consider "good" and a certain set of values that we consider "evil." Where a person falls on this value scale between evil and good determines the entire content of their character. On the rare occasions where Hollywood creates a movie with more moral complexity, there are inevitably protests because people get offended by the idea that their value system might not accurately capture the complexity of life. The primitive Aztecs were so offended by the suggestion that their *historical narrative* might be wrong that they would frequently try to sacrifice anybody who attempted to challenge their narrative. I suppose we truth-seekers should be very happy that in this day and age the bloodthirsty mobs are civilized enough that the only thing they try to sacrifice is people's careers, reputations, and social ties.

The inherent assumptions that our culture makes in terms of its value system of right and wrong is something that I call the *heroic narrative*. Right now, our *heroic narrative* follows a system of guidelines that we call deontological ethics. In other words, some actions are always bad, regardless of the outcomes. For example, killing children is bad, and therefore according to deontological ethics it would be wrong for a random time-traveller (such as chief Montezuma, who keeps wandering into all of my thought-experiment examples) to go back in time and kill Hitler as a child, despite the millions of lives that would be saved. Deontological ethics are in contrast to consequentialist ethics, which simply focus on the end result.

The way that I would best describe the difference between consequentialist ethics and deontological ethics is through this thought experiment: imagine that you were given the option of turning the world into a utopia where nobody would ever die from sickness, assault, or environmental causes – only old age. However, the price of that utopia is that "X" people would die, where "X" is some random number. (That aspect of the thought experiment is the most realistic, since no great change ever comes without **any** death: there are always powerful interests invested in any status quo, no matter how corrupt or inefficient, and historically those elements have not always had the grace to stand aside when change needs to happen - which is why revolutions occur with uncanny frequency throughout history.) If you were a strict deontologist, you would probably not accept such a bargain, since "murder is wrong" is a guideline that falls under the blanket of deontological ethics. By contrast, if you were a consequentialist, your ethical directive would be to always go for the best **end state**. In other words, if failing to accept the "utopia deal" would result in MORE than X people dying from the various causes of death that a utopia would otherwise protect them from, then most consequentialists would say that it would be your ethical duty to accept the deal. Essentially, consequentialism is a more sophisticated version of the trolley problem.

Despite the fact that consequentialist ethics tend to deliver much better results due to the fact that their sole concern is the end state, most people today follow deontologist ethics. Many rationalists have suggested that while consequentialist ethics are desirable, they are difficult to implement because nobody can accurately predict the future. My contention is that this is pure nonsense. In my opinion, we all have a limited ability to predict the future; we simply repress that ability subconsciously because it discomforts our modern deontological value system and leads to disturbing implications about the nature of consciousness, as well as frightening conclusions about the direction that our future is currently headed.

Let me give you an example. The heckler's veto is a long-known principle of group dynamics which states that any system which requires unanimous consent to function will inevitably be paralytic and dysfunctional due to the difficulty of getting unanimous agreement. This is a major reason why many banks and other large institutions are so inefficient – because large deals often need to get people in multiple departments to sign off on them, and many of these people are not incentivized to move the deal forwards. Feedback loops (where good behavior by groups or individuals is rewarded, and bad behavior is punished) are also critical to any organization's success. These two principles of group behavior are generally agreed upon by most rationalists. The EU not only has poor feedback loops, but it is also structured around a system of consensus similar to the heckler's veto. Analyzed from the perspective of an egregore, the current EU system is a lame, sick, pathetic thing that the forces of Darwinism will eventually put out of its misery. Yet there seems to be a striking lack of consensus even among rationalists about the fact that the EU – as it exists in its current state – is doomed to die. Any good rationalist would tell you that these group dynamics are straightforward, and the principles are easy to grasp. The future of the EU is easy to see, it is simply that most people do not **want** to see it because they have an emotional attachment to the EU and the principles of western liberal democracy, so they let their emotions override their logic. Seeing an unpleasant truth and refusing to accept it is called denial, and it is a fundamental flaw that almost all people suffer from.

Another example is border control. We all know that a nation's inability to control its borders carries a cost in human life due to the flow of illegal drugs, human trafficking, and other crime. I call this cost the recurring *death tax*, since it results in an ongoing loss of life each year that the problem persists. We all know robust border control could be easy to implement – all we would have to do is set up landmines at the border, with gaps only at specific highways and checkpoints where border control can do an inspection. This would cause some death when it was first implemented, since there are always a few idiots who will ignore warning signs and walk into danger. However, once the lethality of the system was demonstrated, it seems very unlikely that people would continue to walk into clearly demarcated landmine areas. (And I personally question whether the continued existence of people this stupid would be a net positive for society.) In other words, by instituting these kind of robust border controls, we would be trading a **recurring** *death tax* for a **non-recurring** *death tax*, where all the costs in terms of human life are paid upfront. Most organizations agree that non-recurring costs tend to be better over the long term than recurring costs (in fact, this is a large part of the reason that the company Amazon has been so wildly successful). From the moral perspective of a consequentialist, it makes sense to pick the option that results in the least amount of overall suffering, since that leads to a better end state. So why aren't all borders land-mined? In fact, why is this suggestion so controversial that most governments won't even consider it, or at least do a cost-benefit analysis of which approach would save more lives in aggregate? Frankly, it is because most people in our governments are deontologists, and the idea of little children blowing up on landmines bothers them. On some subconscious level, politicians know that in the long term, many more children will die from a failure to impose border control, due to drugs and human trafficking. Furthermore, these deaths will be truly horrific – deaths from drug abuse, from STDs contracted through sex slavery, etc. However, since those deaths are not **directly** attributable to their choices, our politicians feel no guilt over those deaths, because in a deontological system of ethics, there is no responsibility to minimize overall suffering, or be concerned with the ultimate outcome. For deontologists, any deaths and suffering that are not caused by their direct action but instead by their failure to act are "out of sight, out of mind." To a consequentialist, this way of looking at the world is monstrous. Deontologist ethics cause countless amounts of suffering and – if left to progress to their ultimate conclusion – will eventually result in the complete destruction of our species. (This topic will be discussed more fully in chapter 5.) On the individual level, deontologist ethics are extremely disadvantageous from an evolutionary perspective since they induce an unhealthy state of self-delusion and denial which prevents people from accepting the rationalist principles of group dynamics that would enable us to accurately predict the future.

    It is perhaps unfair to exclusively blame politicians for this, since they are merely reacting to the will of the voters. It is unlikely that many of you would vote for a politician who ran on the "landmine" platform of border control, even if it was mathematically proven to have better outcomes overall. The reason why is because the idea of people blowing up on landmines as a result of your choices makes you feel like a bad person. You want to feel like a hero, and blowing people up on landmines does not feel like a heroic thing to do, even when it saves more lives and spares society from much more suffering than the landmines cause. Hollywood entertainment teaches us that making difficult ethical choices like this is something that heroes never have to do, and a lot of our ethical and moral values have been subconsciously shaped by the psychological conditioning of Hollywood's vast mass media machine. This is something that I call the *heroic narrative*. Every culture throughout history has had different mythologies about what proper heroic behavior is, and these mythologies shape the moral and ethical beliefs of that culture. For better or worse (generally worse), Hollywood shapes our *heroic narrative*. Movies such as Spider-Man or Batman are simply our culture's version of the Odyssey or the Iliad.

So here is my question for you: why are we allowing the most morally and intellectually bankrupt members of our society to shape our view of what proper moral and ethical behavior is? It doesn't take a lot of brains or courage or integrity to be an actor. The key to success in this field is simply to be able to parrot lines, fake emotion, and shamelessly self-promote ourselves. Are these traits something that we want our children to aspire to? Other historical cultures heroic narratives were built by great generals, philosophers, and scientists. Why are we allowing our heroic narrative to be shaped by people like Harvey Weinstein instead of Socrates or Julius Caesar? Do the hypocrites and fools who shape our mythologies have any sort of higher claim to ethical behavior? From an evolutionary perspective, does it seem like a wise strategy to let people like Roman Polanski or Woody Allen shape our image of right and wrong?

In short, what I am trying to say is that you all have the ability to predict the future – you simply make a subconscious decision not to, since the future involves untold suffering and death and it is all your fault due to the system of deontological ethics that you subscribe to. By refusing to accept this truth and living in a state of denial, you manage to avoid the guilt associated with your numerous moral failings. Additionally, you also limit your capacity for self-improvement, since there are many tactical, strategic, and financial advantages to being able to predict the future. The purpose of this book is to strip you of your self-delusion and show you the world as it really is, rather than the pleasant lie that deontologists have created. In other words, we are going to upgrade to a more accurate *historical narrative*.

Chapter 2
~~Augury~~

~~Divination~~

~~Occultism~~

~~Premonition~~

~~Clairvoyance~~

Game Theory

THE ILLUSION OF FREE WILL: HYPNOSIS AND POST-FACTO RATIONALIZATION

      A long time ago, a group of people did a very interesting experiment in hypnosis. This experiment has been replicated many times in many different formats, but the essential details are the same. A person under a hypnotic trance is told that when a certain "trigger" occurs (such as a bell being rung or a phrase being spoken) they should do something ridiculous (such as hopping on one foot in the corner or spinning in circles). The trance is ended, returning the hypnotized subject to full wakefulness with no recollection of the commands that they were given under hypnosis.

      Later on, when the trigger is given, the subject executes their preprogrammed instructions. Nothing unusual there – after all, the point of this experiment is typically to demonstrate the power of hypnotic control. Hypnotists, like all of us, enjoy being given the chance to show off every now and then. However, what I found to be the most interesting part of the experiment is the subject's behavior afterwards. When confronted about why they are doing such a ridiculous thing, they inevitably come up with all sorts of plausible excuses and rationalizations. For example, if the command that they were instructed to execute was to stand in the corner and hop on one foot, the subject might say that they had noticed that part of the floorboard was loose and they were just hopping up and down to test it. If the command was to spin in circles, the subject might claim – quite insistently and passionately – that they saw a wasp buzzing around them at the corner of their eyes. Only when recordings of the entire hypnotic session and aftermath were shown to them would they acknowledge that the entire subjective reality which they had created for themselves **never really existed**. One might say that their hypnosis imbued them with a faulty *historical narrative* that was inferior to the *historical narrative* of the hypnotist, who had a superior perspective thanks to perceiving the entire situation from a better vantage point.

      It may seem silly to you that anybody who knew they were in a hypnosis experiment could do something as outlandish as jumping on one foot in the corner while enthusiastically insisting that this was perfectly reasonable and logical behavior, but consider this: I have been in plenty of conversations on the internet where I knew for a fact that somebody was going to save money on the Trump tax plan because I knew how much money they made and I had done extensive research beforehand on the exact details of how the law impacted their specific tax bracket, but the people on the other side of the discussion insisted that they were not saving any money on the plan at all and that in fact it would hurt them. Further questioning revealed that they didn't even know any of the details of the tax plan – they simply thought it would hurt them because that is what the 30 second soundbites on their favorite news media channels were telling them, and also because this was a deep blue state where it was fashionable to hate President Trump. These people were so clueless that they had not done even the most **basic** levels of research into the topics that they apparently had Very Strong Opinions about. Their behavior had not gone unnoticed by the public, of course – some insightful pundits had already pointed out that the behavior of these people was as repetitive and devoid of self-awareness as the characters in a video game. But where others saw "NPCs," I saw hypnosis. This was an almost exact duplicate of the early hypnotic experiments that I had read about in my childhood. Mass media acted as the hypnotist, programming the subject with all sorts of instructions, like what clothes to buy, what opinions to express, or whom to vote for. Afterwards, the subject would react in the preprogrammed way when they were exposed to the triggers, all while insisting that this was a perfectly reasonable and normal way to behave. And if challenged, they would grow indignant and angry until you shoved the truth right in their face and rubbed their noses in it – again, exactly like the hypnosis experiment. Only after you had shown them the delusionary outlandishness of the lies that they were telling themselves would they sheepishly admit that they had been hypnotized the whole time. "Why didn't I ever think to even question Anderson Cooper?" they might say. "Some random old white dude with nice hair gets on TV, reads something in a tone of somber gravitas, and I just swallow up whatever claim he's dishing up as truth? I don't even spend 5 minutes to go online and **fact-check** him? Seriously, how could I **not** have been hypnotized?" And they would be right to ask themselves these questions. It's not like news media has been such a reliable source of impartial truth or journalistic integrity lately. Why should Andersen Cooper be automatically believed when he is caught exaggerating the intensity of a flood he was reporting live from and responds that he only did so to "be out of the way of rescuers?" I fact-checked his claim with online searches back when that incident occurred (then again while writing this book) and I certainly didn't see any significant rescue operations of any sort in the background, whether pedestrian or automobile. And there certainly weren't any panoramic shots of the area to corroborate his claim – I know because I had actively searched for them. Considering the large number of massive screw-ups that the media has made lately, I don't believe that they deserve to be given the benefit of the doubt in cases like this. Surely, I wasn't the only one with enough critical thought to spend five minutes fact-checking Anderson Cooper instead of accepting his excuse at face value? I've seen people putting themselves literally in a hypnotic trance with help from Instagram, so it seems to me like skepticism should be my default posture, especially when dealing with entertainment media designed to manipulate perceptions.

   Anyway, my point is that if there exist a ton of gullible idiots so credulous that they have a tool as powerful as the internet at their fingertips and they don't even bother to fact-check what a well-paid talking head at a reporter's desk is telling them – well then, perhaps we should have a little more empathy for the people hopping up and down on one foot in the corner while insisting that this is perfectly rational, reasonable behavior. Perhaps **all** of us are hypnotized to a greater or lesser degree throughout every hour of our lives, and we simply aren't aware of it. We rationalize away all the subconscious programming that society instills in us – no matter how unhealthy or ridiculous – as a perfectly reasonable, rational, even logical way to behave. A psychologist might describe this as "an unhealthy state of self-delusion and denial." Think back to last chapter, where I mentioned that deontologist ethics induce an unhealthy state of self-delusion and denial which prevents people from accepting the rationalist principles of group dynamics that would enable us to accurately predict the future. It isn't coincidence that I used this phrase. In my opinion, deontological ethics are basically one of the various forms of hypnosis that are transmitted through mass media entertainment. From a very high-level perspective, you could say that our entire system of deontological ethics is simply an egregore using mass-media as a way to replicate its ideology, thus perpetuating its own existence.

So it seems relevant to talk about how an egregore might perceive the world. In fact, let's talk about the nastiest egregore of all, the egregore which is currently dominant in our world right now - at least, in the current *historical narrative*. Those readers well versed in the writings of the rationalist community will of course know that the egregore whom I am referring to is none other than Moloch.

MEDITATIONS ON MOLOCH

Several years ago, a rationalist of some note who goes by the pseudonym Scott Alexander (it is quite common for members of the Intellectual Dark Web to go by pseudonyms in order to avoid murderously "woke" activist lynch-mobs like the ones that killed Cambrian Genomics) wrote an insightful and popular essay entitled Meditations on Moloch. Understanding the ramifications of this essay is key to understanding how to predict the future and I suspect that even the author himself hasn't fully unpicked the full implications, so I would like to take some time to summarize this brilliant essay before nitpicking some small yet crucial details that I don't think Scott got quite right.

Essentially, Meditations on Moloch is Scott classifying an egregore's strengths and weaknesses through the lens of game theory. The author provides several examples of a game theory problem called the Defector's Dilemma – essentially a structural mental defect that got accidentally hard-coded into us thanks to the evolutionary process that created life. The Defector's Dilemma (which was known to our ancestors as the Tragedy of the Commons) essentially states that for any variety of problem where Life can gain a significant short-term advantage in preserving or replicating its own pattern (typically by acquiring prestige, wealth or power) by betraying its own principles, it will almost always do so, even if this damages its own long-term position. This systemic failure compounds over the course of time until it eventually results in catastrophic disaster. The Defector's Dilemma does not just operate on the individual level, of course; several of Scott's examples show how it also scales up to corporations and nations. For example, most executives of chemical companies don't **want** to pollute the planet, creating a hellish wasteland that our descendants will curse us for, but if they don't take shortcuts in their environmental protections then they will be at a financial disadvantage compared to other companies. Since corporations are egregores and egregores want to survive as long as possible (that being the only basic principle inherent to all life) the executives who constitute the egregore's murmuration will do whatever needs to be done for their company to survive and be competitive, even if that destroys the environment and results in the eventual destruction of their entire species. Moloch is essentially "the egregore of egregores" – the collective embodiment of all this selfish short-term behavior with no consideration for long-term outcomes. In fact, I would speculate that Moloch may even be the reason why we have never encountered alien life, despite the fact that the structure of the universe tells us that it must have evolved countless times in various solar systems. It's not that alien life doesn't emerge; it's just that it keeps evolving to a certain point and then self-destructing due to the Defector's Dilemma. In my opinion, our galaxy is most likely littered with countless amounts of ancient debris from collapsed alien societies which keep missing each other not in space, but in time. Since the evolution of life takes billions of years whereas industrialization tends to change the composition of the environment pretty fast, civilizations that don't resolve the Tragedy of the Commons problem very quickly after hitting full industrialization capacity probably tend to eat dirt pretty hard. And in all fairness, the Tragedy of the Commons puzzle is pretty hard to solve because it's so counterintuitive, so if I were to wildly speculate, I would guess that it probably ends up destroying pretty much **every** alien civilization. In other words, the reason that we haven't encountered other intelligent alien life (despite the astronomically high odds that some must have evolved by now, given the state of our universe) is because 99.9999999 percent evolved right to the point that we are at right now, failed this crucial test, and ended up killing themselves off.

One of the most interesting points that Scott makes in Meditations of Moloch is that people's behavior is always shaped by the incentives of the system within which they operate. That is why structural changes are so difficult to implement. The structure of every institution is composed of processes and the incentives created by those processes, resulting in a certain tendency for the system to perpetuate its patterns. As you will recall from Chapter 1, this is a characteristic of life and the flow of data within that institution is the DNA of an egregore. The flow of information could be described as its nervous system, the flow of money might be described as its respiratory system, the flow of status might be described as its circulatory system, while legal or military enforcement mechanisms could represent the musculature. Unlike me, Scott represents this incentive structure not as the body of a living creature, but rather as topography, with the course of history being represented by a river that flows along the incentive landscape. While I feel that my representation is more accurate and may be more useful to those of a certain mindset, Scott's topographical representation is better suited for people who prefer to generate mathematical visualizations of data. Furthermore, the field of physics already provides us with a set of equations that govern the relationship between water droplets and topography, so all we need to do to test hypotheses is to change some constants and a few symbols, then plug in different values for the quantifiable variables (such as stock market prices, for instance). The usefulness of Scott's model in terms of visualizing and testing experimental data means that in the near term it will undoubtedly become the dominant tool for visualizing outcomes, though in the long term it is important to remember that different *historical narratives* may one day *update* this model. In other words, just because we may visualize data systems as stationary or immobile does not mean that they are harmless or easy to control. Patterns have a life of their own, and sometimes need to be dealt with on their own terms from within the system.

In his Meditations of Moloch essay, Scott evinces a certain sense of despair that even though we may be able to mathematically represent the future as topography, we as individuals are too small and insignificant to be able to effect any significant sort of change to the structure of this topography, since we are effectively trapped within the system ourselves. If I am steel-manning him correctly, Scott seems to be saying that predicting the future is effectively useless if we do not have the power to change it. Considering how much money our intelligence departments have poured into getting just the slightest chance at more accurate forecasting, it would be fair to say that I strongly disagree with Scott's opinion here. If one could codify this super-forecasting knack into a set of quantifiable principles, which could then be refined through further testing and experimentation, then I think it's fair to say that with a little bit of luck, skill, and common sense, the dude who hypothetically managed to pull something like that off could potentially end up becoming literally the wealthiest person on the planet, at which point they would be very well-positioned to **test** whether Scott's assumption that we are too small and insignificant to effect real change to the system actually holds water. I mean, just being able to predict stock markets would basically be an infinite money-making machine. Being able to predict the outcomes of elections would be a source of great political influence. And being able to predict the evolution of our technology tree could really put some kick back into science. There could be a lot of versatility to this "prediction thing" if one were to apply it correctly, which is why I started working on it immediately after reading Scott's essay. For the past several years, I have read a lot of rationalist hypotheses about group dynamics, found ways to measure their hypotheses, and then tested them to see if their hypotheses were accurate or not. What I found surprised me. It turned out that a lot of group behavior was reasonably predictable – not by the laws of sociology (which under testing turned out to be largely false) but by the laws of game theory.

You want to know some people whom I think really understood game theory? The ancient Romans. All of their tactics, their methodology of government, their organization, their use of symbols... it really showed a ruthless efficiency and clarity of planning. In most of the battles they fought, the Romans had a clear tactical advantage, even when opponents had the advantage of numbers. Likewise, the terms of surrender that the Romans offered their enemies were finely optimized to encourage their enemies to surrender rather than fighting to the death, because they left most existing power structures in place when their terms were met, while imposing vindictive punishments upon enemies who insisted on doing things the hard way. In a way, Romans were the ultimate rationalists. They took over enemy tribes, improved them with technology and techniques designed to increase efficiency and *update* their *historical narrative*, and then let them keep doing pretty much whatever those tribes had always done, as long as they didn't try to overrule the law of Rome (which enforced free speech as well as racial, sexual, and religious tolerance). They were good to their allies and vicious to their foes, a behavior which maximally incentivized compliance. Consider that they created one of the greatest empires that the world had ever seen – a feat that many countries today seem incapable of – despite the fact that their industrial base had no advanced machining techniques.

In fact, I admire the Romans grasp of game theory (an ancient science that the modern world - deluded by its own sense of self-importance – has long forgotten) so very much that I would like to tell you a parable about them. The purpose of this parable is to illustrate the inevitability of certain historical events, when considered in a different scope.

The Parable of Marcus the Plebian

Marcus walked through the woods into town. It was one of those small towns in Dacia that would eventually crumble into dust, as all things do, but at that moment in his life he thought it quite bustling and prosperous. Though he did not know it then, his remote town was living through what would later come to be known as "the Fall of the Roman Empire." Attached to his belt were two things: his purse, and his sword. He needed both in the execution of his duties, for Marcus was a tax collector. The utility of the purse was to transport the money citizens owed to the State, and the utility of the sword was to defend the purse, since large quantities of money always attracted scavengers. His father (a tax collector in his own right) had told Marcus stories of a time when such guarded vigilance was not necessary, since men feared the power of the State far more than any sword, but these were the times and so these were the precautions.

A shriek from the woods startled him, though he maintained his composure. It was two of the Goths, of course – lurking in the woods waiting for locals to spook for their own malicious amusement. The payments to the Goths were late, due to the economic crisis in Rome, so as the Goths numbers grew, they increasingly started to see the native Romans as targets, since the natives' lives were far more prosperous thanks to the civilizing influence of the State. It is important to remember that the payments were not TRIBUTE, never tribute – technically the Goths were refugees so Marcus had been instructed to refer to the payments as "development funds." Development funds that the mighty Roman empire chose to bestow upon the poor unfortunate Goths; that was the appropriate narrative. It would not do to question why, if the Roman empire was so "mighty," the citizens of their border towns lived in fear of the Goth refugees, who frequently robbed their houses, groped their women, and soiled their public commons. After all, if one questioned the ability of the State to enforce its laws upon a handful of ragged barbarians, then people might question the legitimacy of the Empire itself, and that was unthinkable – it implied that the gods had withdrawn their ancient blessing from Rome, as some heretical seers claimed. So the payments were technically not tribute, they were "development funds" to elevate and enlighten the poor benighted savages. Unfortunately the Goths did not seem to have been apprised of the Empire's politically expedient nomenclature or narrative since they certainly ACTED for all intents and purposes like an invading horde whose tribute had not been paid. In other words, they got drunk and restless, stole things, harassed people, and generally caused havoc until enough soldiers could be raised to send a military expedition to fuck them up.

Marcus had worked for some of these military forces in his younger days, perhaps driven by some deep-rooted desire to see the world before settling down in Dacia. He had encountered trash like this and knew their behavior well. If he had been weak or vulnerable, the scavengers might have attacked, sensing too little risk to outweigh the potential rewards of his purse. But an obvious veteran of the Legion would clearly have the skill to kill at least one if not both of them, so they hesitated, their greed and recklessness slowly fluctuating but never outweighing their fear. Marcus drew his sword and watched them calmly, observing the delicate dance of emotions as they played across the two Goths faces. He had spent time training at the Temple of Mars, so he knew exactly what the outcome of this encounter would be even before the Goths did. They would slowly and painfully do the math, coming to the conclusion that Marcus was too hard a target to overwhelm by force. To salvage their injured pride, they would attempt to startle or frighten him, perhaps pretending to charge and laughing when he shifted into a fighting stance. Then they would slink off into the woods, walking slowly in order to save face and make it clear that they were not running away. He drew his sword, allowed the encounter to play out, and then continued on his path. A few miles down the road, he came across a dead swallow lying on its back across a shallow crack in the road, its wings outstretched in the shape of the Roman Aquila. Marcus, who understood omens as the parables that they were, took note of this carefully.

Thessaly answered the door when he finally got to Valerian's sumptuous home. "Is the governor in, Tess?" Marcus inquired. "I have the monthly tax payment." The slave girl rolled her eyes. "He's... occupied. But you're welcome to wait." She didn't have to say more. Everyone knew that the governor didn't like to spend much time on actual governance. "I can have the kitchen staff fetch you some pork stew." "I'd appreciate that," Marcus said. "The roads are dangerous at this hour; I'd prefer not to travel them with a full purse." "Those damn Goths," Thessaly sighed. "I'll be honest with you, Marcus – lately I've felt unsafe every time I step outside the walls of this home." The two of them made friendly conversation for a while. Eventually they were interrupted by Valerian emerging from his bedroom, a prostitute reclining on the bed behind the unlatched bedroom door.

"My dear Marcus! Reliable as ever... if a few days late." "I couldn't travel by night," Marcus said by way of explanation. "Too risky with all the Goths." "Yes, it's a sad situation," Valerian said. "But what can we do? The government simply can't afford the development funds." He stretched out his hand for the purse, and Marcus handed it to him slowly, though not without a glance around the lavishness of his surroundings - the better to see where the wealth of the State was going. The governor emptied the purse upon the hard wooden table and began counting, sorting the coins based on their mint varieties.

"Perhaps we should roust the Goths until they start behaving more like a conquered people than the conquerors," Marcus suggested casually. "With what funds? Do you have a treasure trove of dragon's teeth I don't know about? Because the State simply cannot afford another military campaign." Valerian finished counting. "Well done, Marcus. It's good to know that I can still rely on you... unlike my other tax collectors." "The State's other tax collectors," Marcus corrected him. He had always been a quick thinker, but now his mind felt like it was moving so fast that it had become unmoored. "Yes, that's what I meant. Between you and I, those lazy slackers have been running very light lately. They claim that the Goth invasion makes it too dangerous to venture into certain areas to collect taxes, or that they were robbed on the road – it's just one excuse after another with these people. That's what I like about you Marcus. You guard the State's purse well."

"That is the duty of the sword, after all," Marcus said, and promptly used his sword to do so. Blood spurted out over Valerian's clothing as he slumped backwards. There was a sharp intake of breath as Tess stopped herself from screaming so as not to raise the prostitute from her drug-induced stupor. Valerian looked up at Marcus accusingly. "You traitorous dog," he said. "Of all my servants, I never thought you would be one to betray the State." "The State is dead," Marcus answered simply, and as he spoke the words, he could FEEL their truth. The omen that he had seen made sense to him now. "It died a long time ago, and we have simply been playing charades with its corpse. But there will be a new State someday, and if it rises in my time, I will continue to serve it."

"You won't get far. The guard will find you, hunt you down." Valerian coughed up blood.

"No, they won't," Marcus said. "If they still had the power to enforce law and order within this dominion, they would be doing so upon the Goths."

"May the gods damn you, Marcus," Valerian gasped, and then died. "No, Valerian," Marcus replied. "The gods have blessed me. With insight." He bent down to pick up the scattered coinage, and there was a soft click as Tess shut the bedroom door. "Are you coming?" he asked.

"To where? You're a fugitive, and I'm a slave." Tess looked so shocked that Marco had to laugh. "Damnit, this is not funny, Marcus. You're a wanted man now. I'll pretend that I saw nothing, but Valerian was right – the guard will eventually be after you."

"Tess, you don't understand. You still think of yourself as a slave. That was the old world, where the State still existed. This is the new world. There is no State. No more than fifty miles travel from here, there are towns which do not care about the empire. There will be nobody who knows if you are a slave or a queen."

"And what will we do for a living, Marcus? How will we eat? You may have a sword but you're too old to be a soldier."

Marcus smiled. "Don't you see, Tess? The State has fallen. The barbarians have already taken over Dacia – maybe not in name, but in action. They may be fierce fighters but they have no grasp of tactics. The techniques of the Temple are a mystery to them. With the State gone, those techniques are priceless. And the techniques of the Temple are my new stock in trade."

And then Tess smiled too, because she had finally begun to understand.

THE STRUCTURE OF INCENTIVES

What are the lessons that we can take from Marcus's parable? For starters, please take note of the short-term perspectives of most of the characters in this story. We as readers have the superior *historical narrative* because we are examining this situation with the benefit of many centuries of hindsight, so we understand that all of the major events described – the increasing quantity of Goth refugees, the breakdown of law and order, and economic depression – were all symptoms of a collapsing Roman empire. But nobody in this story could see that systemic problem, because they didn't **want** to see it. The idea that the State might totally collapse was unthinkable to them, so no matter how strongly the evidence pointed in that direction, they simply deleted that idea from their minds. In an earlier chapter, I wrote that "Seeing an unpleasant truth and refusing to accept it is called denial, and it is a fundamental flaw that almost all people suffer from." All of the characters in this story suffer from exactly this set of mental blinders. It is only when Marcus learns to identify and terminate his own personal self-delusion (assisted by his training at the Temple as well as a useful omen) that he starts to interpret the events in his life through the lens of our superior *historical narrative*, at which point he can see the future of the empire with ease, and takes action to ensure himself a significantly advantageous position within that future. Where would Marcus have ended up if the gods hadn't blessed him with that insight? Most likely, the roads would have become increasingly unsafe as law and order continued to break down in the outskirts of the Roman empire. The Goth refugees would multiply until instead of outnumbering tax collectors two to one, it was three to one. Or ten to one. Or twenty to one. Eventually they would have stabbed Marcus to death out in the wilderness somewhere due to the increasingly tempting target that his purse presented - because that is the way large groups of people **always** behave, no matter how virtuously they may present their motives to themselves and to others. When their fear outweighs their greed, they do not attack. When their greed outweighs their fear, they do. Large groups of people behave much like bacteria in that regard, which is unsurprising since this fundamental fight-or-flight principle is one of the most basic imperatives of all life. Life wants to engage in behaviors (ie, status-seeking, acquisition of wealth and power, victories over would-be rivals, etc) that help replicate its pattern most effectively, and Life wants to avoid Death. This means that you can shape the behavior and structure of large groups in exactly the same way as you can shape the behavior and structure of bacteria in agar dishes, simply by applying positive feedback (incentives) and negative feedback (punishments) to exactly the right spots.

Anyway, the lesson I took from this story is that we may never know what ultimately happened to Marcus (since parables generally don't extend beyond the central moral point that they are making) but at least we know he didn't end up with the unhappy ending that would doubtless have befallen him if he had kept his mental blinders on. Personally, I like to think that Marcus had a very long and prosperous existence. He seems like the kind of guy who is generally pretty good at landing on his feet.

WE DON'T BUILD THE SYSTEM; THE SYSTEM BUILDS US

Now that we have a certain degree of insight into the structure of incentives and how they can shape the behavior of large groups over time, let's analyze some other large groups to see how the incentive structure that was imposed on them led to suboptimal outcomes and problematic behavior. This can help us codify some of the principles of incentive structures to help forecast how large organizations behave. These principles are *useful* because if you can predict what incentives lead to suboptimal outcomes, you can easily find ways to attain more optimal outcomes simply through the process of elimination. Quite simply, when you filter the garbage out of data, what remains is valuable.

Let me give you an example. What shapes our patterns of behavior? If we are operating under the *historical narrative* that we think things and then do things based on those thoughts, then presumably it is our belief system. But if we are operating under the *historical narrative* that ideas are like a murmuration of starlings that nests in our heads, then our patterns of behavior are simply the manifestations of how our mental patterns are shaped by the data that enters us on a day to day basis. Change the data that people receive, and you can change what they think, which modifies their behavior. In terms of social structures, this is reflected by the way that our behavior is guided by the positive and negative stimuli that we receive throughout the day in terms of feedback from other people. An unkind way of phrasing this might be that we are literally sleepwalking through life, doing things largely for the sake of how other people perceive us. This is demonstrably true from the fact that anonymous donations to charity are miniscule in proportion to public donations. If we were doing nice things just for the sake of doing them – rather than factoring in how others perceive us – then there would not be such a large discrepancy. The only logical conclusion is that there is a lot of social signaling involved in our charitable contributions. In fact, my experimentation has led me to conclude that there is actually quite a lot of subconscious social signaling going on "under the hood" in literally everything we do.

On the bright side, this means that we can shape other people's behavior through a process called operant conditioning. During a brief phase in my life where some might have unfairly called me a pick-up artist, I had the opportunity to test this in a rather amusing manner at a nightclub where I would often meet women. (This was back in the long-long ago, the ancient time period before Tinder - a barbaric and uncivilized era where the only way for men to meet prospective mates was to actually go out in public and strike up conversations with strangers.)

Since this was a nightclub where people tended to be less inhibited, other pick up artists would often operate there as well, like wild animals returning to the watering hole. Being a bit of an insecure twerp at that time in my life, I took an eventual dislike to one of them, a tall and handsome gentleman who was a flamboyant dancer. I had recently read an amusing anecdote about how the founder and creator of operant conditioning, B.F. Skinner, had a group of his psychology students secretly use his own techniques against him, conditioning him to deliver his lectures from the doorway to the lecture hall auditorium, with one of his feet out the door. (Incidentally, in reading that anecdote, please note the strong similarities that operant conditioning has to hypnosis.) I decided that regardless of whether the anecdote was true, I should test out whether such subconscious conditioning was possible – partially because it is always *useful* to test out the reality of your own paradigm to ensure you are operating in the most modern *historical narrative*, and partially because it would be absolutely hilarious to see this dude who was typically very successful at picking up women get inexplicably shut down for an evening and have no idea why. Some rationalists may take our science quite seriously, but I'm not one of them. What can I say? I like to have fun with it.

Did I even have a plan? Or did I just want to see if emotions were contagious in large groups of people? Either way, I soon had my answer. Dancing near him, I evoked an expression of disgust in his direction, angling my shoulders so as to face away from him, and attempting to shoot him contemptuous micro-expressions. Shortly thereafter, I saw other people mirroring me subtly. Women were turning their backs to him, and giving him the kind of look typically reserved for creepy stalkers. The most interesting thing is that the process started very slowly, but picked up speed as the emotional "virus" spread. It seemed like there was a direct mathematical correlation between the number of people whose micro-expressions radiated contempt towards the pick-up artist and the speed at which other people adopted the same emotion. The crowd was literally reacting like a flock of sheep, all mimicking behavior without any conscious understanding of why they were doing it. To me, this was the proof of concept for a discipline that I call memetics. We will discuss memetics more in chapter 3, but it is impossible to understand the full scope of game theory without at least a brief description of memetics, because each is the yin to each other's yang. **Game theory** studies the flow of ideas within groups, and **memetics** studies the manipulation of those ideas. We utilize both disciplines subconsciously on a day to day level – for example, game theory helps us model other people's behavior through a process called "empathy" while memetics is used to influence other people's behavior through a process that some would call "charisma." However, because this is all done on a subconscious level, we are unable to interrogate our own behavior deeply enough to quantify and improve it.

    The point of this anecdote is that the process of deliberately crafting an emotion to be transmitted through a group really awakened me to how much of our feelings and beliefs are created by the social systems we live in as well as the people around us. One example of this would be our ethical beliefs. When we are incentivized to have certain beliefs and opinions out of our own naked self-interest, we have those beliefs but then rationalize more upstanding "moral" reasons for holding those beliefs. For example, you want other people to think well of you, so you donate to charity. If you acknowledged to yourself that the reason you are donating to charity is mostly social signaling, then you would be forced to think of yourself as a bad person. So you rationalize your virtue signaling by telling yourself that you donate to charity because you are a good person… but you make sure to leave your name on the check and post about your donation on Facebook. That's why public donations far outweigh anonymous donations. On some level, you're aware of the true motivations behind your behavior, but you just don't let yourself think about it because under our current system of deontological ethics, that would make you a bad person. In other words, what we perceive of as our "selves" or our consciousness – what psychologists would call our "ego" – is 99% self-delusional bullshit.

    So here is the idea that I am putting forwards: why allow ourselves to be brainwashed by all this deontological ethics nonsense? From a utilitarian perspective, the only societal purpose of deontological ethics seems to be advancing the self-interest of the elite 1%, because it hypnotizes the masses into forgetting that if the 1% ever exploit the rest of us too much, we can simply choose to cooperate with each other to kill them - with no more justification needed than that their continued survival is not in our self-interest. And since one major side-effect of this mass hypnosis is that it prevents us from having the self-awareness necessary to understand why we do the things that we do, it makes it really hard for us to notice when we fall into self-destructive patterns. Maybe instead of repeating the same tired old patterns that turn us into brainwashed slaves, we should try something a little bit new. But in order to do that, we need to examine the patterns of the past to see how those patterns self-perpetuate. After all, it is only through studying parasitic egregores that we may learn how to defeat them.

HOW GARBAGE PATTERNS REPLICATE

As our parable's protagonist Marcus tried to point out to Valerian, patterns of behavior replicate when they are given incentives to do so, and die out when they are given disincentives. When the amount of incentives to do something outweighs the disincentives, people magically begin behaving in that way, and rationalize the reasoning for their behavior. The Goth refugees behaved in an aggressive way because it was profitable for them to rob and sexually harass people, and because Rome was too weak to punish them when they behaved in ways that promoted civil disorder. I am sure that the Goths created some convenient rationale in their minds for their vicious behavior such as it being retaliation for Roman "colonialism" - just as many Romans undoubtedly rationalized their own empire's weakness as "charity" - but at the end of the day, none of these superficial reasons really explains their behavior. The truth is that every party in this parable simply acted according to their short-term incentives and came up with a convenient fiction in their minds to justify the morality of said actions, much as a person acting under the influence of a hypnotic command will come up with creative ways to rationalize their own behavior, no matter how delusional it may be. Likewise, when the structure of incentives changes in such a way that different patterns of behavior become more rewarding, people's behavior changes to reflect that. Marcus was loyal to the remnants of the Roman Empire until he saw that continued loyalty would only lead him to death while betrayal would lead to greatness, so he changed his behavior accordingly. Was Marcus's betrayal blameworthy, or was it the fault of the Empire for failing to create a social structure that would reward loyalty? Were the Goths bad people, or was their behavior the fault of the Empire for failing to create a social structure that would adequately punish illegal behavior? Personally, I don't think it is helpful to assign blame for people's behavior since ultimately our behaviors are all products of our environment. Through the mechanism of government we collectively create the social incentives that shape societies into heavens or hells, and then we are forced to live in the social structures that we created, reaping our just rewards (or punishments). Let's examine some of those structures through the lens of game theory, analyzing their strengths and weaknesses as egregores. This will let us understand how to develop egregores that are healthier for us and more conducive to our long-term prosperity.

The three examples I would like to discuss are the Clinton campaign, the Antifa movement, and Islam. These three examples are interesting because even though two of these social structures are political and the third social structure is religious, they all share the same weakness – a susceptibility to doctrinal hijacking. This weakness stems in all cases from a feedback loop between two fundamental forces that play an important role in group dynamics – the *Virtue Signaling Escalation* and the *Expanding Circle of Retribution*. I'll define both of these forces so that we can understand why the interaction between them created such a fatal weakness for these egregores.

Virtue Signaling Escalation is the principle that when people are divided into an ingroup and an outgroup and there is no penalty for doing bad things to the outgroup, people will attempt to gain higher social status within their ingroup by saying and doing increasingly bad things to the outgroup until it eventually leads to attempted war or genocide of the outgroup. For example, persecution of the Jews during World War 2 didn't start immediately with death camps. It started with people saying "Look at those privileged wealthy elitists. Always using nepotism to help each other get ahead, while the rest of us have to struggle down here in a system that's rigged against us." While the long-term social engineering implications of having an extremely insular religion can be discussed at a later time, the most important point here is that people had enough genuine resentment built up towards the Jews that when one German publicly exclaimed "Fuck the Jews" and the rest of their tribe agreed, that person gained social status. Remember that Life instinctively seeks out situations that allow itself to replicate (at least, in the short term) and social status is one of those things. Having a high social status allows people to more easily gain reproductive opportunities and money – in other words, Life wants to replicate its patterns (mates are attracted to high-status people) and Life wants to avoid Death (money can buy personal security). So in a way, we might say that the Holocaust was inevitable from that first angry yell. Once that first angry person in the mob yelled "Fuck the Jews" and gained social status instead of reprobation, then other people were incentivized to do exactly the same thing. However, since the shocking taboo had already been broken, screaming out "Fuck the Jews" again would no longer have a psychological effect as strong as the first time. In other words, once attacks on the outgroup become normalized, there is a diminishing return on social status, which requires subsequent attacks on that group to be more severe in order to achieve the same yield in social status. For example, the next person would have to yell "We should make them pay for what they've done!" in order to achieve the same return on social status as the first person. Then the person after that would have to escalate still further in order to achieve the same social status yield; "We should beat them up!" Then it would escalate to property damage, then physical damage, then government-sanctioned violence, etc. The same principle occurred during the Communist revolution in Russia, except that the privileged elite being targeted there were the nobility. Understand that there is always a privileged elite for less fortunate people to blame their problems on (sometimes legitimately, sometimes not) and the cycle of violence always starts exactly the same way – with a group of people that for whatever reason it is considered socially acceptable to express hateful feelings about. Once the outgroup has been established as a legitimate target for hate, the hateful feelings **always** inevitably morph into violence because people subconsciously crave the social status that they gain from condemning the outgroup, so they rationalize reasons for doing so. This is what I call "doctrinal hijacking." It does not matter how peaceful or well-intentioned the in-group originally intended to be: over time, the structural incentives of the system gradually turn them into a bloodthirsty mob of violent savages. To be fair, some outgroups (such as those who judge others on the content of their skin color rather than their souls) may legitimately deserve to be condemned. But where incompetent social scientists focus on questions of right and wrong, talented social scientists prefer to study how the process works in order to better understand it, since it's hard to maintain objectivity when you have an emotional stake in the experiment.

The *Expanding Circle of Retribution* is a built-in defense mechanism that we evolved specifically to protect ourselves against the *Virtue Signaling Escalation*. Remember that we were all born with the ability to predict the future, and it is only the mass hypnosis of deontological ethics that limits our ability in this regard. This means that on a subconscious level, we are all aware that a single angry yell can be the catalyst that eventually leads to violent lynch mobs. Naturally, people in general tend to get very nervous about what look like tribally sanctioned attacks against them. So nervous in fact that when a member of a different tribe attacks our tribe, either through slanderous rhetoric or action, we have the reasonable expectation that the group that they belong to ought to punish the offender, in order to show that they do not stand behind their actions. Failure to socially punish the offender (or even worse, rewarding them with social approval) is typically considered a declaration of war because the other tribe is essentially saying that they condone the attacks against you or even consider such attacks laudable. And like all group conflicts, these things unfold in very predictable patterns of escalation.

All of the above is a complicated way to explain why – on the individual level – we have evolved a defensive behavior against a process that we are not fully perceptive of on the conscious level. Of course, the explanation is only complicated if you accept the *historical narrative* that our belief systems originate inside our minds. If we prefer to think of our ideologies as egregores, always fighting each other in a Darwinian competition of tooth and claw, then it makes perfect sense that they would develop natural defenses against each other. It is as natural a process as any other kind of evolution.

In any case, the best way to explain how the *Expanding Circle of Retribution* works is to understand the evolutionary forces which caused these memetic behaviors to form. The first point that I need to make is that vengeance is not a maladaptive trait. On the contrary, the desire for vengeance is **highly** adaptive for survival. Every cooperative species needs a mechanism to defend itself against defectors (by this I mean the "game theory" version of defectors), and the easiest way to defend oneself against a threat is to destroy it. In fact, I would go so far as to say that it is the **optimal** way. Even chimpanzees and birds do it to members of their own species who are not team players. Don't you know why they call it a murder of crows? It's because crows periodically kill members of their flock who commit crimes. This leads to increased teamwork, cooperation, and coordination.

The *Expanding Circle of Retribution* is the mechanism by which vengeance is subconsciously routed through the mind. This mechanism is the reason for social dysfunctions such as familial blood feuds, gang warfare, vendettas, and other such charming quirks. Essentially, the stream of thought – our subconscious programming at the evolutionary level - operates according to the following simple program, which we then rationalize in various unusual ways.

10. An good person is wronged by an evil person. I don't really believe in objective good or evil, but from an evolutionary perspective the best way to define a good person would be "Somebody who operates according to the shared collaborative laws that mutually benefit everybody" and the best way to define an evil person would be "Somebody who does not operate according to those shared collaborative laws." Please note that I am

specifically limiting this definition to shared laws that benefit and constrain everyone in a roughly egalitarian manner. Laws that do not abide by that principle are both evil and stupid from an evolutionary purpose, because it is detrimental to our long-term well-being to give up any of our freedoms without some corresponding guarantees of health, wealth, comfort, or safety that adequately compensates us for sacrificing the prerogative of exercising those freedoms.

20. The desire for vengeance is triggered. For that desire to go away, vengeance against the evil person must be achieved in some form. This is a highly evolutionary behavior from a group perspective because it reduces the incentive for people to become narcissistic sociopaths, and obviously tribes die from massive internal schisms when they become too full of self-obsessed people, which is why it is *useful* to gradually whittle down their numbers, or at least discourage them from replicating their negative patterns of behavior. (Please note that I am drawing a clear distinction here between self-obsession and pride – it is justifiable to take pride in your accomplishments only if you have actually accomplished things of **note**. As long as we are speaking in purely evolutionary terms, I think it is fair to say that neither your race, gender, or what you "identify" as count as accomplishments.)

30. In the old days when we lived in an honor-based society, this would be the point in time where the good person would pick up a heavy object of some kind and bash the evil person's head in. However, in the current age we live in a rules-based society rather than an honor-based society, which means that people are not allowed to independently act on their natural vengeance instinct. Instead the good person's grievance must be escalated to an authority figure – somebody within the evil person's tribe who outranks the evil person.

40. If said authority figure does not redress the grievance adequately by punishing the original defector, the need for vengeance remains, except now the authority figure is included in the desire for vengeance because they sided with the evil offender. This is why I call it the *Expanding Circle of Retribution*. Regardless of their tribe, authority needs to act from a neutral perspective in punishing wrongs done to us, otherwise there is no evolutionary incentive for any of us to preserve that authority. On the contrary, it is vastly to our mutual advantage to destroy evil authority (ie, authorities that do not adequately protect our individual interests). In other words, evolution programs us to desire vengeance against things that have hurt us because those things are detrimental to our well-being. Authority that cannot be counted on to represent us fairly certainly counts as "detrimental to our well-being", so that is why the circle of "legitimate vengeance targets" expands.

50. Because we live in a rules-based society, one cannot just take out the authority figure, so they typically escalate the appeal for just redress of the injury (ie, vengeance) to somebody one level above THAT authority. Predictably, it is ignored, since partisan tribalism has infected everything, and by the time you get two levels above the person who has wronged you, it's a little unrealistic to expect them to resolve the problem by punishing several people from their tribe simply on your say-so.

60. The good person looking for justice/vengeance (remember that justice and vengeance are exactly the same thing from an evolutionary perspective) gets increasingly frustrated at the lack of punishment for the evil defectors. It is important to society to

punish people who defect, since a stable civilization depends on internal cohesion. If you cannot count on authority to prevent others from unjustly attacking you, then why not exploit the system yourself by attacking anyone you please? Only an idiot would obey rules when the other side does not.

Here is an example of the Expanding Circle of Retribution in a low-tech society, for example the semi-tribal villages of the Ural mountains.

10. Somebody shoves somebody else and refuses to apologize because they are an asshole. It escalates to a fistfight.

20. The person who was shoved seeks vengeance.

30. Because they want to respect the rules, they can't simply imbed an axe into the person who shoved them. Instead, they take it to the head of the family of the person who shoved them and plead their case, explaining why they are entitled to retribution.

40. Since the person they are pleading their case to is related to the shover, they are biased towards them and more inclined to take their side. They refuse to punish the shover who triggered the fistfight.

50. The person takes their case to the magistrate, who blows it off because they have more important things to do.

60. With no suitable punishment for the defector, the injured party feels that the law has failed them and so they are now entitled to ignore it. They get a bunch of people together to beat up the original person who shoved them. Predictably, things get more and more violent, due to the *Virtue Signalling Escalation* process, until somebody gets badly hurt.

Outcome: BLOOD FEUD! In the Ural mountains, these can last entire generations.

Of course, this only happens in primitive societies. In modern societies, our tribes are larger. Instead of being shoved, the "wrong" that people typically suffer is something like being unfairly banned on twitter. Have you noticed how, in all social-media "moderation," left-wing people are allowed to say vicious things about the right while the right is often censored and banned from the discourse when they try to respond in kind? The same thing applies to minorities speaking in a derogatory way about the majority, and the majority being unable to retaliate because of the "punching down vs punching up" doctrine. The **reason** all of our currently existing social media companies have structured the discourse in this unnatural way is because they are all exclusively based out of Silicon Valley, a hyper-liberal echo chamber with some extremely idealistic yet naïvely delusional views about the design of human nature. The Left justifies this algorithmic censorship with some nonsense about "power plus privilege" but regardless of whether you believe in that principle or not, the fact is that evolution has not wired us to behave in such an unnatural way, and handing out platitudes about society when vengeance is expected is not a great way to "read the room."

Because the majority of Twitter employees are liberal, they tend to ignore conservative complaints while favoring liberal ones. Essentially liberals get away without punishment in situations where an impartial observer would say that their punishment is most certainly called for. Over the course of time (and repeated iterations which incentivize this behavior) this lack of punishment results in assholes gravitating to the Blue team.

Red tribe members seeing this get angry and escalate their demands for redress (without response, since the left will not punish other leftists). Since the "proper channels" do not result in the required response to eliminate their need for vengeance, they eventually route through "unconventional channels." This is how we get votes for conservative candidates who promise to punish the left. The funny thing is that it is not really the conservative aspect that people are voting for - it is the punishment. We need to accept that the desire for vengeance is a fundamental part of human nature, instead of repressing and demonizing it. As I have said many times throughout this book, it is unrealistic to expect people to act in evolutionarily maladaptive ways. It doesn't matter how many books about critical gender theory or whatever other academic nonsense one quotes to try to sublimate this aspect of human nature - the fact is that people who are unfairly treated want vengeance, and if you try to mess with this process, the process ends up messing with you.

So essentially, Twitter censorship is an industrial strength engine for producing microaggressions against anybody who doesn't agree with the liberal viewpoint. Each act of unjust censorship or stifling of a viewpoint can be considered one microaggression. Since it is *inevitable* that at some point, any given individual will disagree with the liberal viewpoint, repeated microaggressions trigger the same neural pathways, reinforcing them. This eventually results in a deep-seated bloodlust against the agents - and the *system* - responsible for those microaggression. Individually each microaggression is insignificant, but because of the incentive structure that social media censorship creates, the overall impact is very significant. We have already seen this happening in terms of the Republicans gradually increasing hold on power.

So now that we understand the underlying principles behind the *virtue signaling escalation* and the *expanding circle of retribution*, let's go back to the three examples I mentioned earlier so that we can examine how this subconscious programming plays out in the real world.

Within Islam (at least in terms of how it is practiced in many Middle Eastern societies) the "ingroup" is one's fellow believers and the "outgroup" is western civilization. It is cool and fashionable for many people in the Middle East to badmouth the West. (In other words, the status boost acts as an incentive.) There is no penalty or punishment for doing so. In terms of their behavior, large groups of human beings behave in exactly the same way as bacteria – they automatically gravitate towards the behaviors that gain them some reward, unless there is a penalty for doing so whose severity outweighs the reward. For bacteria, the reward is nutritious agar solution, and for human beings the reward is social status. For bacteria, the punishment is bleach, and for human beings the punishment is social stigma. While the final location within the Petri Dish of any individual bacterium is hard to predict, the ultimate form that the entire bacteria mass will grow into can be predicted perfectly if you understand where the nutrients are and where the bleach is – to the point that we can easily create art with it. Similarly, if you understand the underlying structure of incentives and disincentives present in a system, you may not be able to determine the specific fate of one individual human, but you can still perfectly predict the shape that their **society** will grow into.

Since Islamic practitioners (again, as the religion is commonly practiced in the Middle East) gain a social status reward for badmouthing the U.S.A. and there is no punishment for doing so, they will badmouth the U.S.A. There is nothing mysterious there.

Because there is no penalty for verbally attacking the outgroup (in this case, the USA) the *Virtue Signalling Escalation* principle states that the hostility of their behavior will escalate over time, since maintaining the same level of hostility results in a lower status boost, necessitating increasingly aggressive behavior in order to get the same relative "dose" of social status. (If this sounds similar to the description of a narcissist seeking their "narcissistic fix," that is by design – evolution is fundamentally narcissistic in nature.) Eventually hostility escalates to the point that a terrorist group forms.

The USA registers a complaint of some sort with the leaders of the society in question, but the complaint does not result in any meaningful change, because it is much more convenient for the leaders of those societies to be able to blame "Western Colonialism" for the problems within their own societies rather than have their citizens grapple with the idea that said problems might actually stem from their own leaders mismanagement.

The USA then escalates the complaint to the UN and other global organizations, but nothing gets done because much like the Roman Empire during its decline, these multipolar organizations have devolved into ineffectual bureaucracies in perpetual deadlock. It takes a truly momentous event to get the UN to do anything significant, and even then, the effort is usually spearheaded by the USA. In short, trying to get the UN (or its economic sister, the WTO) to punish individual nations like China for misbehavior is like trying to get your driver's license at the RMV without any identification. It just is **not** happening: not because those organizations are malicious, but because the people within them are incompetent and so those organizations have lost any spark of vitality that they may once have had.

The cycle continues, and people in the USA get frustrated enough that nothing is done to punish individual Islamic radicals that they expand their anger to include the leaders of the Islamic societies in question (for failing to stop the radicalization of their own society) as well as the UN (for failing to punish the Islamic societies in question). Again, this is all perfectly predictable thanks to our understanding of the *expanding circle of retribution*. Eventually, public frustration within the USA builds to the point where voters start electing politicians who campaign explicitly on the promise of punishing Islamic society and withdrawing support from multipolar global organizations such as NATO or the UN. Again, I want to repeat that this is **not** unhealthy behavior, nor is it in any way maladaptive. Authority that cannot be depended on to protect our interests is worse than useless and from an evolutionary perspective we collectively ought to destroy such authority whenever we encounter it, because it is disadvantageous for us to trade away the freedom to do whatever we want without gaining adequate privileges and protections in return. Another way to say this would be that we are strongly disincentivized from doing so, and just as the rest of society acts according to the incentives acting upon them, so too do we. The fact that we rationalists study the interplay of forces that act upon society does not in any way render us immune to those forces' effects.

Antifa has exactly the same problem as Islam – it is an outrage culture based around escalating hostility to an outgroup (in this case, conservatives) with no built-in disincentives for individual misbehavior. Because individual Antifa members gain higher ingroup status by badmouthing Trump supporters, and there is no punishment for their hostility, then the *Virtue Signalling Escalation* eventually leads Antifa into a spiral of increasingly violent behavior against anybody whom they perceive as "fascist," a definition which seems to constantly expand in scope.

Conservatives attacked by Antifa attempt to gain the vengeance/justice that they are fairly entitled to by escalating to legal or political authorities, but this seldom has an impact, because the cities that Antifa is primarily based in tend to have liberal politicians, which means that there is no social or political will to clamp down on misbehavior among their own tribe. This failure of local authority to represent conservative interests fairly means that the targets of conservative anger expand to include the liberal politicians and cities who are failing to adequately discipline bad behavior within their own tribe, and eventually they elect politicians who campaign on a platform that promises (either openly or through tacit dog whistles) to punish such bad behavior. Moderate centrists who are worried about the constantly expanding scope of whom Antifa defines as a "fascist" tend to throw in with the conservatives, helping them gain the votes needed to win. Again, this is all perfectly foreseeable thanks to the *Expanding Circle of Retribution* principle.

    The Clinton campaign followed the exact same pattern of *Virtue Signalling Escalation*, with increasingly aggressive rhetoric against conservatives that eventually reached the point of self-parody when the presidential candidate herself made a series of extremely serious accusations against a cartoon frog. What distinguishes this example from the others is the fact that after their defeat, Democrats made accusations that Russian spies had used propaganda to influence the election. We may never know the truth about such rumors, but in theory it is quite possible to achieve such results by using undercover agent provocateurs to act as a catalyst for the Virtue Signalling Process, causing one tribe to radicalize faster. Such rapid radicalization would appear to the outside world as complete derangement, which would certainly example why Hillary Clinton's blue wall collapsed like wet toilet paper. In chapter 4, we discuss techniques to gain such electoral advantages, as well as the basic countermeasures needed to make campaigns less permeable to such tactics. It is worth noting that Hillary's campaign took absolutely none of the recommended countermeasures that a competent memetic strategist would consider simple due diligence.

    The point I am trying to make here is that human patterns of behavior are highly predictable. With the correct model, every one of these outcomes could have been anticipated beforehand. The exact methodology of manipulating group behavior is covered in more detail in the next chapter, which focuses on memetics. However, before we can move on to manipulating group behavior it is important to be fully and completely aware of just how much of our group behavior is subconsciously predetermined by evolutionary psychology. The answer is "pretty much all of it." This is *useful* because being able to predict group behavior allows you to exploit situations where knowing the future could be advantageous, such as political events and the stock market.

MAPPING MORALITY WITH ETHICAL CALCULUS

The most basic defining characteristic of all Life is that it moves towards incentives (ie, things that enhance Life's ability to replicate its pattern) while avoiding disincentives (ie, things that increase the chances of meeting Death). This is useful because it allows us to model incentive structures with a sort of topographical map that allows us to predict the future with a high degree of accuracy.

Imagine the life of every human being to be a drop of water flowing along this topographical map. Incentives can be mapped as depressions in the topography. Disincentives can be mapped as elevations in the topography. The reason we map things this way is because the closest analogous model to group behavior is fluid dynamics. Just as water flows downhill and tends to settle at the lowest available point in the area, so too do people gravitate towards incentives, and tend to settle at the static equilibrium that is best for them, in terms of what they can achieve within the confines of the system.

For example, imagine a busy street with two sidewalks on either side. The sidewalk on one side of the street is cracked and dimly lit. It looks like the kind of place where a person might get mugged, or where they might stumble and twist their ankle. The sidewalk on the other side of the street is brightly lit, and has several eye-catching storefronts. On a regular basis, these stores offer free samples of their products and have visible signage to indicate that. What could we expect to see in terms of the group dynamics of people walking down the street? Scientific testing has indicated that over time, the side of the street which looks safer and occasionally offers free promotions will see marginally higher foot traffic, as people gravitate over time towards rewards and away from hazards. In terms of our topographical map of incentives, we might say that the dimly-lit side of the street has a slightly higher elevation than the brightly-lit side of the street. If the collective "flow" of foot traffic were to be represented as a river (with a single water droplet representing the movement of an individual person) then the majority of the water's flow would travel along the lower elevations. Please note that I am drawing an important distinction between "majority of the flow" and "all of the flow." There are always going to be outliers - people who are not familiar with that specific location, or who don't think it's worth the time crossing the street to be on the more brightly-lit side, or who simply prefer darkness. This means a fluid dynamic model seems to be relatively consistent in terms of predicting the **probability** of any one particular person/water droplet travelling/flowing along a particular trajectory/incentive path.

Now's let talk about slope. In the above example, the incentives to walk on one side of the street and avoid the other side are fairly low, so we would say that the slope of the topography is fairly shallow. People will slightly favor the more brightly lit side of the street, but not disproportionately so. For example, if we monitored the movement of people with no external preferences for favoring one side of the street over the other, we might potentially expect 55 percent of them to favor the more brightly lit side, and 45 percent to favor the dimly lit and cracked side of the street (in much the same way that a swiftly flowing fluid running through a shallow channel would "favor" the side of the channel that is deeper).

Now imagine that on the brightly lit side of the street there is an altruistic rationalist giving out ten-dollar bills to everybody who passes by, while on the dimly lit side of the street there is a deranged looking vagabond muttering loudly to himself while visibly brandishing a knife at passersby. We could reasonably expect that in this situation, a **lot** more people would prefer to traverse the brightly lit side of the street that had a free money giveaway to the dimly lit side of the street with a potentially violent lunatic. If we monitored the movement of people traversing the street under these circumstances, we would see approximately 95 percent of people travelling along the brightly lit side of the street with only 5 percent of people (predominantly thrill seekers and those too caught up in their cell phones to notice threat) travelling along the more dimly lit side of the street. In terms of our incentive topography, we would say that the dimly lit side of the street now has a much higher elevation than it did in our previous example, while the brightly lit side of the street now has a much lower elevation. We would thus describe the topography of the incentive flow as having a steeper slope, because it funnels crowd behavior more effectively in one direction, much as a swiftly flowing fluid running through a channel with a deep groove on one side would have the majority of the fluid passing through the groove rather than splashing along the other parts of the channel.

By the way, according to American Psychological Association guidelines, this is a completely unethical experiment. This is why sociologists are so inept at understanding human nature; because they are legally prohibited from creating experiments that would reveal it. To be honest, I have discovered that modern sociology is a garbage field with almost no redeeming value, because it is impossible to get a truly accurate understanding of how people behave when the subjects are aware that they are being observed as part of an experiment. Fortunately, Dark Rationalists understand that the pursuit of knowledge is a sacred vocation that sometimes requires unconventional approaches. So the next time you see a filthy looking homeless person acting crazy, remember that appearances can be deceiving: you may in fact be participating in an interactive experiment conducted by the next generation of aspiring Dark Rationalists.

Getting back to the point, what this topographical analysis means is that not only can we **predict** group behavior with the correct map, but we can also **manipulate** the behavior of large groups of people by giving them the correct incentives. Even a small change in the incentive topography can be extremely significant in situations like presidential elections, where a slight majority of voters may be enough to irrevocably alter the course of the future. In chapter 3, we focus on general guidelines for manipulating large groups. Specific techniques for rigging elections to favor your preferred candidate are discussed in chapter 4 (as well as best practices for defending your campaign from these kinds of manipulation). For now, it suffices to say that a lot of the structural problems in our society are caused by the fact that our current system of deontological ethics tends to encourage compassion rather than punishment towards people who behave in a way that is detrimental to society. Since slope is defined as the difference in topographical height between incentives and disincentives, such a huge bias towards compassion makes it very difficult to manipulate society in a way that generates good outcomes. If you have ever wondered why the UN and EU are such ineffectual organizations, this would be your answer. In fact, I would say that the existence of both of these organizations is actively harmful to the world because of their insistence on an outdated code of ethics which may **seem** well intentioned (by deontological standards), but which results in massive amounts of suffering and death in the long term. If you want to have better long-term outcomes from a utilitarian perspective, you need to be fully prepared to renounce compassion and act in a vindictive way when the incentive topography calls for it.

Let's use the Catholic Church as an example, since they tend to operate in "compassion failure mode" very frequently. I'm sure we've all heard of the widespread pedophilia scandal that is currently still ongoing within the Church. Many of us were shocked and dismayed to hear that priests were regularly raping their parishioners' children, but the fact that I personally found most shocking was that this abuse turned out to be systemic. In other words, through the use of incentive topography one could have perfectly predicted that this child abuse would happen, because by definition a systemic problem is one caused by the incentive structure of a system. To put it in layman's terms, pedos like to diddle kids, right? Whether this is due to deliberate malice or some old neurological relic from the days when human life expectancy was 18 is irrelevant; the point is that it functions as an incentive for them. For the purposes of incentive topography, you could represent "job with access to kids" as an indentation on the topographical map of pedophile incentives. (Please note that since pedos have different incentives than your average person, their topographical map will look different.)

It stands to reason that in order to counteract the gravitational effect that such a lower elevation would have on the "flow" of pedophile behavior (in other words, the **position itself** would draw people who are motivated by such an incentive structure, just as water flows downhill) you would need deterrents of equal or greater strength to elevate the topographical terrain enough to stop it from creating such a moral sinkhole. In other words, vigilance (to maximize the chance of such behavior being caught) and harsh punishment, should a priest be caught molesting children. A bit of proactive foresight might be useful as well. For example, if I were the Pope, I would have spyware installed on all of my employee's computers to monitor their porn usage (assuming that I even had a significant IT department at all), with a data-scraping algorithm that would disregard normal porn while raising red flags if a priest was caught repeatedly viewing scantily clad minors below the legal age of consent. Anybody who triggered too many flags would be considered a potential problem and psychological counsellors could be brought in before the priest's behavior escalated. Even setting aside the fact that algorithmic analysis is *useful* for being able to anticipate and curb deviant behavior among the Catholic Church's employees, the Church could better serve its flock by using more modern methods to minister to them, rather than operating in a paradigm thirty years behind the times. Considering that one of the main roles of religious figures in ancient cultures was therapeutic, it is a little puzzling that priests nowadays are not trained in one of the most useful tools of psychological profiling. Even a quick look at a parishioner's Instagram feed or Reddit participation would be enough to yield a treasure trove of information about them. (Chapter 4 covers this technique in greater detail.) Some might argue that parishioners would be hesitant to let priests have visibility to the most personal and vulnerable details of their lives to analyze in a non-judgmental way, but I was under the impression that this was kind of the whole point of the confessional. You see, if I were the

Pope, I would not be running my business as casually as if I had locked my users into a 2000 year old exclusive monopoly. I would be actively trying to find ways to serve my customer base more efficiently. Considering that these people literally **want** the Church to make a difference in their lives and most priests **want** to make a positive impact in their community, you would think that this would be a match made in Heaven. So why is religious belief currently at all-time lows? It's possible that updating one's historical narrative is *useful* for success, whether you define success as the business of making money or saving souls.

    From a certain point of view, there is not that vast a difference between "optimizing a society's future outcomes" and "building a better world," nor between "using algorithms to map the structure of a person's incentives" and "analyzing a person's soul." So perhaps using the Catholic Church as an example of what we could achieve through incentive topography might actually have been a much more appropriate use case than I originally intended.

Chapter 3
~~Bewitchment~~
~~Enchantment~~
~~Mesmerism~~
~~Charm~~
~~Bedazzlement~~
Memetics

THE GAME

In the previous two chapters, we discussed how most group behavior is actually very predictable through the lens of evolutionary psychology (or "game theory" as it is more commonly known, for political correctness reasons) and how our entire belief system of deontological ethics is nothing more than a system of mass hypnosis that allows us to persuade ourselves that our choices are the result of free will, despite the fact that we are effectively operating in a sleepwalking state. This explanation might be very unsettling to those of a thoughtful disposition, because it raises a terrifying question: if we are all subconsciously obeying the directives of some ancient evolutionary program while simply rationalizing said choices, could we possibly be **re**programmed? And how hard would it be to execute said reprogramming? I'm afraid I have some bad news for you. We are all being reprogrammed every day of our lives by the things we read, the media we watch, and the people we interact with. Most of these reprogramming is done accidentally, without strategy or purpose. But this reprogramming can also be done deliberately also, and in the hands of an expert the results can be fairly spectacular.

But to get a sense of how memetics works, we need to understand how the study of this long-lost science was reborn into the current era. So let us travel back to the early 2000s, when the science of mental reprogramming was still in its infancy. This is because it didn't have a lot of funding, due to the fact that it was primarily practiced by young semi-autistic men with deep self-confidence issues. Wait a second, that describes almost all hard sciences. Perhaps the drive to push science forwards may in some deeply subconscious way stem from the fact that no matter how big their brains are, men seek to gain prestige because it maximizes their access to mating opportunities. They will do this using every advantage at their disposal (such as superior wealth, charisma, or intellect) because Life wants more of itself, and Life wants to avoid Death. Gaining prestige in the tribe satisfies both of those goals. Of course, this is just speculation. I haven't done an in-depth study of every single STEM field in existence to validate my hypothesis. However, in the case of memetics, I can confirm without doubt that the early adopters of this science were trying to get their rocks off, because the initiates in question were pick-up artists and during the time period in question, the Venn Diagram of "pick-up artists" and "memetic researchers" formed almost an entirely perfect circle. Let me put it this way: there was a lot of free-floating testosterone floating around those rooms. Speaking from an incentives perspective, you could say that these young researchers had a lot of skin in the game.

The way these pick-up artists operated was basically by reducing human behavior to a program template similar to a Choose-Your-Own-Adventure story. It was depressingly simple. The PUA (Pick Up Artist) in question would spot an attractive woman and deliver a line that she was not quite able to pattern match. Because the mind evolved as a pattern-matching tool, she would continue making conversation in an attempt to match the pattern. Eventually the conversation would last so long that the woman would convince herself – through cognitive dissonance – that the reason she had spent so long talking to the PUA was because she was attracted to him... and interestingly enough this would **make him more attractive**, because at our core we are whom we identify as, and a large part of our identity is built through the things that we do. In other words, if you think you are attracted to somebody, then you **will** start being attracted to them, even if you were not before. It may sound odd but I have tested this countless times and it definitely holds true.

As with all things, terms and conditions did apply. I noticed that particularly bright women seemed much less susceptible to this effect, and their responses were much more self-aware. For example, a neuroscience student whom I briefly hooked up with once told me about an odd incident that had occurred to her earlier in the day. She had been shopping for shoes when a gentleman approached her and made an unusual comment about her appearance. She brushed him off dismissively, and then was approached by another gentleman asking exactly the same question. And then another. All in all, she was approached by five gentlemen making the same kind of comment. What an unusual day, she commented.

What I didn't say to my lovely fling was that the particular area where she had been shopping was almost next door to a place where a notorious pick up artist was giving a seminar to several hundred young men on how to "improve their game." It may seem hard to believe, but at the time there was a huge market for that skillset because this was before the widespread use of online dating. I recognized the shoe line as a classic pick up artist move because it was specifically designed to blur pattern recognition. There was something existentially depressing about the way pick up artistry reduced people into a series of buttons to push in order to achieve the desired results, so I recall being very pleased that my date was too smart to fall for a canned line. I was determined to look into it further.

Over much time and experimentation, I would learn that there was indeed an inverse correlation between a person's self-awareness and their susceptibility to this technique that many had begun to refer to as neuro-linguistic programming. In other words, the more self-aware a person was, the more difficult it was to reshape their conception of themselves. Ironically enough, it was this discovery that led me to abandon the pick-up community, since I found the idea of dating women with low self-awareness very unsettling. Perhaps it was just wild paranoia, but I felt like I could never be entirely sure how much of their personality was genuine and how much they had subconsciously interpellated from me.

NEURO-LINGUISTIC PROGRAMMING

Of all the Dark Arts, I think that neuro-linguistic programming is the one which is most overhyped. Radical proponents of NLP have made absurd claims such as claiming that it can cure disease or enable people to contact God. Meanwhile, pick-up artists trying to sell "seduction classes" have pushed the unscientific (and unrealistic) idea that neuro-linguistic programming can make attractive women immediately willing and compliant for carnal bliss. In my opinion, it is most likely these wild exaggerations that have led the discipline into such disrepute. That said, it is still important to understand NLP as a foundation for memetics, so we need to separate which aspects of neuro-linguistic programming are fact and which are myth.

Myth: Neuro-linguistic programming can make people do things that they don't want to do. This is a misconception which arose from the fact that people impacted by neuro-linguistic programming often make wildly unconventional choices that at first glance do not seem to serve their best interests. For example, it is a stereotype of pick-up artists that they often date women who far outclass them both professionally and in general terms of having their shit together, so friends of those women often choose to perceive them as "mind-controlled" because it is easier than trying to understand something that their *historical narrative* offers them no frame of reference for – namely, the fact that different people may look at the same person and see different qualities in them.

Fact: Neuro-linguistic programming can change what people want to do. Well, sure – this is a no-brainer. People change their minds about what they want to do literally all the time. In the USA, there even exists a strange caste of people whose entire role in society is getting people to change their minds. The best way to think about this is as a change of perception, or an *update* of one's own internal narrative. For example, some people reading this book may think I am a delusional crackpot, particularly if they have invested a lot of their life and self-worth into the outdated *historical narrative* that I am debunking. Others may have made a lot of money off of the principles that I describe here, and so that group of people would probably be more likely to perceive me as a visionary scientist whose unconventional approach resulted in his exile from the ranks of the ignorant elites who dwell in academia's crumbling halls. A handful of people may even have convinced themselves somehow that my unconventional use of rationalist techniques is supernatural in nature, and that group of people might potentially view me as either a divine prophet or a witch. Alternatively, if this last group of people were more scientifically oriented, they might instead perceive these writings as the work of an alien intelligence. Each of these groups of people would obviously react to me in a different way depending on how they perceive me. This does not mean that any of these people have been mind-controlled; it simply means that they each have a different frame of reference with regards to me and so everybody makes their best guess about who I am based on the information available to them at that given moment in time. By using neuro-linguistic programming to alter other people's frame of reference, we can change what they want to do. This is because our minds and emotions evolved purely to assist Life in making choices that maximize Life's ability to fulfill its core programming, so we are constantly recalibrating our wants and needs as more information becomes available to us. To that extent, the fact that we are able to change other people's minds by providing them with new information is a feature, not a bug. I would be far more concerned by somebody who was **not** able to change their mind in response to new information since that would indicate that their mind was not functioning the way that evolution intended and it would almost certainly be a sign of deep mental illness.

Myth: Neuro-linguistic programming can radically alter people's core programming. This is false – our core programming is very difficult to change. For example, an ambitious person will always be ambitious. At different points in their life, said ambition might focus on different things: for instance, a person whose ambition is to succeed in the real estate market may one day shift gears and decide to succeed at politics, but their core personality trait remains the same – it is simply redirected in a different way. It is very important to distinguish between changing one's core programming and changing their personality. Core programming determines whom people are at the core of their being, whereas personality reflects how people express their core programming to the world. To use an example from my personal life, I have been in love with different people over the course of my life, many of whom were wildly different from each other. Does that mean that I was a different person at each stage of my life? In the metaphorical sense I suppose we could make a case for that, but in the more literal sense I am still the same person that I used to be, and I simply changed my mind about what I wanted as I became more mature and my tastes changed. Nothing can alter your core programming short of an Act of God, which obviously falls outside the scope of this book. Personality, on the other hand, is far more malleable.

Fact: Neuro-linguistic programming can radically alter people's personalities. One of the most important concepts that I have tried to hammer home again and again in this book is the fact that 99% of our choices are not really choices at all since they are made by our core programming rather than our executive functions. Because most of what we interpret as our own personality is just surface level rationalization of what our deep program tells us to do, obviously adjusting how our programming interfaces with the rest of the world is not as hard as many psychologists want you to believe. (In fact, I often suspect that a lot of the problems with psychology arise from the fact that incompetent psychologists who are unable to solve their client's problems rack up more billable hours than highly competent psychologists who are able to address the majority of their client's problems in just 4 or 5 sessions. If society did a better job of incentivizing psychologists by rewarding competence and punishing incompetence, it's possible that the field of psychology would improve greatly.)

It may be helpful at this point in time to clarify what I mean when I contrast "core programming" with "how that programming interfaces with the rest of the world." For all intents and purposes, almost nobody else in life will ever care what your core programming is: they will only care how it interfaces with them. For example, my core programming tells me that it is enjoyable to occasionally participate in hegemonic systems of patriarchal oppression. Since this is core programming, I can't change that instinct, but I can easily change how this instinct interfaces with the rest of society. For example, if I were a bad person, I might have expressed that instinct by becoming a perv. Since I am a good person, I chose instead to dabble in alternative lifestyles and had some very interesting experiences with partners who enjoy this kind of thing as much as I do. Instincts cannot be created or destroyed through neuro-linguistic programming, but they can easily be channeled, redirected, or sublimated into various forms, depending on the goals of the programmer. One important takeaway from this is that since the topographical map of our incentives (which we refer to by shorthand as "instinct") is impossible to change, you should never feel bad or self-critical about your instincts, regardless of what our sex-hating media tells you. Instead, you should work to channel your innate drives in healthier and more pro-social ways. Much of the reason that our modern society is so overmedicated and neurotic is because we have some fundamental misconceptions about how human psychology works, due to the complete and utter failure of modern psychology "experts." These misconceptions stop us from self-actualizing in healthy ways. (From a societal perspective, the fact that politicians and governments regularly use the advice of these ignorant "experts" to generate dysfunctional public policy is even more unhealthy, but that will be discussed more fully in Chapter 5.)

Myth: Neuro-linguistic programming can let you talk to God. Do I even need to address why this is silly? Unless you literally see people performing supernatural feats, you should be very skeptical of anybody who claims to have talked to God. I believe that the reason this misconception arose is because most people have a hard time understanding the full scope of existence, so when their blinders fall away and they see something that falls outside the boundaries of their experience, it shocks them out of their complacency so much that it may easily feel like a religious epiphany. For example, when Native Americans first saw Europeans riding horses, it was so unusual to them that they initially failed to distinguish the horse and rider as being separate entities and instead perceived them as a single being, like a centaur. Due to the way that our minds process information, any experience that falls completely "out of scope" with previously established contextual parameters will usually be perceived in a religious way. This is because the brain is a pattern-matching device, so when we encounter something that doesn't fit any known pattern, our minds tend to glitch until we have fully assimilated the new phenomena into our worldview.

Fact: You can do some really wild stuff to your own mind with neuro-linguistic programming. While you are having fun reprogramming other people's personalities, you may eventually decide to spend some of your time more constructively to do some clean-up of your own internal processes in order to make yourself more efficient, intelligent, and successful. And I suppose if you're using psychedelics to facilitate the process, it might from some perspectives be interpreted as a conversation with God or some angelic entity. Better be careful about what you say to it! "God" might not like it if you cop an attitude, and as this book demonstrates, just because an egregore has no physical existence doesn't mean that it can't hurt you very badly if it wants to. You don't get any sort of diplomatic immunity from your own ideas – in fact, psychiatrists offices are often full of people who have been badly damaged by their minds. So if you take this approach, tread carefully, because you can literally go insane this way, and there is no guarantee at all that you will ever be able to find your way back.

So now that we have established what neurolinguistic programming is and what it is not, how does it work? Put simply, neurolinguistic programming takes advantage of hidden cognitive biases in our deep programming that we are unaware of. For example, pick-up artists frequently use our pattern-matching glitch to induce attraction by making it difficult for the target to pattern-match their behavior. In a similar vein, if you are repeatedly exposed to pictures of somebody whom you have no preexisting feelings about, you will tend to find them more attractive because our brains "prefer" patterns that are easy to match. This is why Owen Wilson is subjectively perceived as an attractive person even though he is objectively unattractive according to most scientifically measurable criteria. It is because of the fact that his repeated exposure to audiences through the medium of film and television makes it easier for them to pattern match him and thus makes him more attractive. (Please note that from everything that I have read about Owen Wilson he is a super nice guy and my observation should not be taken as a personal criticism of him.

Owen, if you're reading this, I hope you're not offended.) A more tragic version of this pattern-matching phenomenon is that people who have been in unhealthy relationships tend to subconsciously recreate those relationship dynamics because the brain has an incredibly strong preference for pattern matching, even when the pattern is unhealthy. The purpose of these examples is to give you a general sense of how hidden cognitive biases in our deep programming work. Chapter 4 contains more practical step-by-step instructions for how other people's cognitive biases can be exploited to achieve a wide variety of interesting results.

  At its core, the "science" of neurolinguistic programming is nothing more than the acknowledgement that most of the powerful drives which motivate us are purely subconscious, and what we perceive of as our "ego" or "self" is nothing more than the collaboration – or competition - of these powerful drives. Unfortunately, it is very difficult to be completely objective or analytical about these subconscious drives for the same reason that it is difficult for a computer program to analyze its own source code while said source code is currently running. Some rationalists have achieved a limited measure of success by using psychedelics in conjunction with NLP to split off a portion of their ego so that they may observe their own behavior objectively, but this technique is not recommended because it often carries severe side effects. A better approach is to observe and analyze your own behavior after the fact to determine what underlying programming may have been responsible for your own actions.

HOW TO SHAPE THE SYSTEM INSTEAD OF LETTING IT SHAPE YOU

  So far, all we have discussed is the application of these principles to creating an idea that can reshape a single individual's personality. While this is *useful* as proof of concept, one of the main purposes of this book is to teach techniques of generating and manipulating large-scale effects. Essentially, what we want to do is give an idea that we are creating the spark of life so that it can go out into the world and act **independently** to shape multiple people's realities. That means that our idea must also possess the characteristics of life – namely, it must be both *predictable* and *self-replicating*, optimally even *adaptive* to its environment. While biologists may argue whether an idea – something that is ultimately nothing more than a collection of data - could really be alive, I would argue that any idea that can live in multiple people's consciousnesses, that can adapt and evolve to fit its situation, that can act on its own to manipulate its environment to its own advantage – well, that certainly seems to fit all the criteria of "life" to me. As the old saying goes: if it looks like a duck, walks like a duck, and quacks like a duck – it's probably a duck. An A.I. researcher might argue that it could actually be some weird artificial duck that has no true self-awareness, a philosophical duck zombie if you will, but I would argue back that if its "simulated" behavior is close enough an illusion to fool 99% of the public into perceiving those actions as conscious behavior, then it's light years more self-aware than your average Tinder date or Redditor. And how sure can you be that **you** possess self-awareness? I mean, we just spent all of last chapter discussing how as societies our collective future behavior is ultimately every bit as predictable as a series of physics formulae, so from where I'm sitting that doesn't look a whole lot like self-awareness to me.

Hopefully these points address my hypothetical A.I. researcher's hypothetical query. On that note, we return to discussing how to make the idea that we are building into a living creature – an egregore. Life is characterized by being both *predictable* and *self-replicating*, so how can we give our newborn idea both of those traits?

We'll start with the trait of predictability. To illustrate this, I'd like to tell you another parable, this time from the point of view of a different narrator. Let's examine the world as it might potentially have appeared through the lens of the world-famous inventor Nikola Tesla.

The Parable of Nikola Tesla

"I don't know why you'd want to see the place where it happened," the servant said. "The body's been cleaned up already. Gruesome stuff."

"I have a scientific curiosity in such things," answered Nikola. "Indulge me, please."

The servant nodded and led him around to the front door of the home, looking around carefully first to make sure that they were not being observed by any of the neighbors. Nikola might have overpaid him a bit to gain access to the house - what was the going rate to bribe a domestic these days anyway? - but this was important. He had to know.

"Here," the servant said. "This is where it happened."

Nikola examined the doorway. The doormat had already been replaced, and for good reason - he had looked in their trash and found the old doormat, which was completely unsalvageable. Liquefied flesh tended to leave stains.

"So he was standing right here when the spontaneous combustion incident happened? Or did he reach out to touch the doorknob?" Nikola asked quietly. "Think back carefully please. This is very important."

"Now you come to mention it, he was definitely touching the doorknob. Strange thing too, he wouldn't let go of it, even when he was all on fire. Gruesome stuff," the servant repeated again.

Nikola opened the door gingerly and examined the other side of the doorknob. A slight discoloration of rust - that might have been the contact point for the wire, which would probably have sparked a bit on the outside. There was a lot of foliage nearby lining the fence. If the killers had a spool, they could have simply spooled the wire back through the foliage as soon as they heard the body drop. If anybody noticed a wire hidden in the shadows of the shrubbery, they would have thought nothing of it, and why would they? Nobody knew how dangerous high-voltage alternating currents could be except for Nikola and a handful of other scientists. Naturally the uneducated and less intelligent would attribute the spectacular death to spontaneous human combustion, and why wouldn't they? Spontaneous human combustion was something that they knew - something which fit into their narrative. In an earlier era, they would probably have attributed it to magic. And in an era before that, they would have said it was the work of the gods.

"Thank you," Nikola said. "I've seen enough."

As he walked home, his mind raced, repeating the same inner monologue that had inspired him to come here - the inner monologue that he had hoped to disprove. You started this, his inner voice said. Once you created this technology, you had to know that somebody would find a way to weaponize it.

But there was another voice inside Nikola also, the voice that had always seemed to guide him when he was visualizing his greatest inventions. Yes, I knew that there would be a little bloodshed, it replied. What of it? That is the nature of humanity itself. Every time you give them something new, they inevitably use it to hurt each other. Does that mean we should stop inventing altogether? Just leave the world stuck in this backwards dark age, with no understanding at all of how reality works? Because if that's the price we need to pay to make sure people don't abuse power, it's not worth it. You want to make us feel guilty about a few broken eggs, but you sure used to love a good omelet once upon a time, didn't you?

At first Nikola had been frightened of that voice, but over time he had grown to love it. It calmed him and reassured his doubts. I promised you that we would build a better world, it said. But I never promised you that the path would be easy. It would be delusional to think that ANY great change to society would ever be easy. And we're not delusional, are we?

"No we are not," Nikola whispered under his breath, and chuckled. No delusions here, just an unnaturally gifted inventor having a private conversation with his inner voice. He had always known that there would be trade-offs. Any time the Great Game changed, there were always winners and losers. The future was like a train, moving forward with an inevitable momentum that was impossible to stop. You could be on the train or under it, but it was inadvisable to get in the train's way.

But he didn't stop to reflect on this line of thought for very long. Thinking about trains had given him an idea. Nikola pulled out the small notebook he carried with him everywhere and started to make notes.

Let's take a moment to put this parable in the context of its time. While electricity was known of in Tesla's day, historically it was viewed in a very different way than it is today. Direct current was the main form of electricity, which meant that resistance attenuated the electrical current and prevented it from travelling far. This meant that until alternating current became more common, all electrical generators had to be local – the idea of an "electrical substation" that could convert all electricity into a more easily transmissible form was considered crazy science-fiction talk.

Of course, we all know what happened next from the perspective of our superior *historical narrative*. This crazy science-fiction idea that everybody had laughed at turned out to actually be real, completely changing the course of history and reshaping society. And humans being what we are, we inevitably weaponized it and eventually turned this gift into something that could be used on each other. From what I have read, Tesla was by no means a stupid man, so it's very likely that he understood that this type of weaponization was inevitable and was OK with it, or at least understood that advancing the future of humanity required some tradeoffs.

Stop! Let's interrupt our reading here to do a little pop quiz, just to see if you have started to think like a proper Dark Rationalist yet. Out of the four sentences in the preceding paragraph, which one is the most important and relevant to the parable?

The answer, of course, is sentence three – "And humans being what we are, we inevitably weaponized it and eventually turned this gift into something that could be used on each other." The reason why that sentence is the most important is because of the word inevitably. If something is inevitable, that means that it is *predictable*, which is one of the two characteristics that we wanted to imbue into our idea to give it the spark of Life. If you prefer to dismiss the concept of egregores as esoteric mysticism and instead subscribe to the more "scientific" sounding incentive structure topography model, then another way to think about it is that if we want our memetic creation to spread, then it needs in some way to be *useful* to the targets. Weapons will always be *useful*, since Life wants to replicate its pattern while avoiding Death (and weapons help with that by providing sharp incentives to other potentially uncooperative Life), so if you want to ensure that your idea takes root – if you want to absolutely **guarantee** it – just make sure that your idea can in some way be used as a weapon, and present it as a gift with no strings attached. If one person doesn't want to grab it, someone else quickly will simply to avoid allowing their enemies to get their hands on it.

Of course, weaponization is only one possibility: there are plenty of ways to package an idea to make it more *useful* to its audience. The broader point I am making is that you want your manufactured idea to be like a virus, but a beneficial one – something that increases the resilience of its host. If you can inject that level of craftsmanship into the process, then it is absolutely inevitable that your idea will spread, even if you do nothing more than whisper it to a single stranger in a darkened room. Some might even say that our own personalities are nothing more than different topographical maps of Life's eternally *useful* ideas, such as Aggression, Desire, Prudence, Dominance, etc.

PREFERENCE CASCADES

We have discussed how to make our idea predictable, but on its own, that's no big deal. Rocks are predictable also. What distinguishes life from unlife is that life reproduces its own pattern – in other words, Life is self-replicating. So how can we make our idea self-replicating? Preferably, we want something that self-replicates fast, because the quicker our idea takes root in people's minds, the more powerful it will be. And let's be real here – you're not trying to create an egregore just because you want an exotic pet, right? You want a powerful tool, and the fact that your tool might potentially be alive is a secondary consideration.

When you're trying to learn something, it pays to take advice from the best. So I think that perhaps the most *useful* way to teach this lesson is through another parable, this one narrated by a man who used ideas like a finely honed rapier to carve out a legacy for himself.

The Parable of Edward Teach

Once upon a time there was a man named Edward Teach, which was appropriate since his life offered so many teachable moments. In a future era he would be better remembered as the infamous pirate Blackbeard, (and have his good name slandered by historians who insisted on spelling it as "Thatch" instead of "Teach") but at the time of our story he was just a desperate man staring Death in the face. You see, this was early in Teach's career as a professional pirate captain, and his inexperience combined with his notorious greed had led him to, in the modern parlance "write some Letters of Exchequer that his sword arm couldn't cash."

Even back in those days, it wasn't a normal thing for mutinies to occur, so the fact that a full two-thirds of his crew wanted to kill him was a clear sign to our little Eddie that he had only gotten to this point through some bad life choices. It was common in those days for pirate crews to be bound by a charter known as the Articles of Association. This document – the pirate equivalent of a legal contract – dictated the by-laws under which the crew would operate and the loot would be shared. It was this latter issue which was causing the dispute, for the Articles of Association of this particular crew gave Captain Teach a much larger portion of the loot than his subordinates were comfortable with. Pirate law was a bit more rough-and-tumble than common law, so litigation over this particular contract was traditionally carried out by blade and bullet. Given that the disgruntled crew members outnumbered those loyal to Teach by a ratio of two-to-one, things looked grim. Eddie knew he had only enough time to say a couple of sentences before the shooting would start.

In moments like this, it was important for a man to make his peace with God, so even though time was short, Eddie closed his eyes and said a brief prayer to the only angel he had ever believed in – the Angel of Rationality. We may never know precisely what words passed between them, or whether the Angel existed only in Eddie's mind, but when he opened his eyes again Blackbeard saw the problem through a new perspective. A more **rationalist** perspective, shall we say. And suddenly, his situation didn't seem quite so insurmountable.

"My lads," he exclaimed, seizing his moment to speak. "You say I've done you wrong through my greed, and I do admit that this was true. The Articles of Association that I initially drew up were unfairly apportioned, I will concede that. But I will concede one thing to you also, if you would have it: a promise." He used his cutlass to scratch a thin line upon the wooden deck in front of him. "This line demarks where my allies stand and where my enemies do. To those who partner with me and stand on my side of the line, we shall share our profits alike, equals in all things," He paused, letting these words sink in. Bear in mind that this was a much darker time, where the signature of a contract was typically forced over a gangplank. In other words, in those days the crew never got an equal cut of the booty. So when Blackbeard offered to treat his crew like equal partners, this was a BIG deal. The other thing to keep in mind is that Edward Teach was a really smart guy, and he was a prodigy when it came to piracy. Often the crew didn't take even a single casualty when raiding, since they would just pull up alongside an enemy boat, bombard them with grenades until the line of defense was broken (Teach was truly a man ahead of his time when it came to deploying the latest and greatest military tactics), and then the other side would surrender pretty quickly. The crew knew Teach was a highly competent leader, and they liked having easy victories – something that was exceptionally rare in the days of piracy. So when you factored that in with a promise to split the booty equally... well, it really gave that crew something to think about. The words writhed in the minds of the crew, quickly burrowing in and taking root.

Blackbeard continued "But to those on the other side of the line – those who stand against me - may they burn eternally in hell, and all my true partners gun them down like the scurvy dogs they are!" At this point, a great many things all started to happen very quickly, all in a very predictable way.

First, the crew already close to Blackbeard's side of the line dove across it to join him, because they saw where this was going and wanted to be on the winning team. They knew they would get a much better deal on his side than they could get from the other side, and they knew that the row of people behind them would (for the most part) make the same calculations. Hey, they were pirates, they tended to be self-interested, alright? Although some might say that all Life is self-interested in its own way, and that it is only those with a surfeit of material possessions who can afford the LUXURY of virtue, or at least to confer upon their own identities the illusion of such.

Second, Blackbeard's crew opened fire upon the mutineers. They were surprised to be fired upon because they hadn't seen their front ranks disappear as they immediately dove across the line, defecting to join Blackbeard. However, each individuals mind made exactly the same calculation about whether to defect and roll across that line, except that now it had to make that calculation under the pressure of live fire. But at the end of the day, they came to the same conclusion – their side was losing people a lot faster than the opposite side. Most of them made the same choice, and the few that didn't were wiped out by focused fire from Blackbeard's troops, leaving the ranks behind them to make the same decision under pressure.

This weighing of the odds – this preference for Life to avoid Death – cascaded down through each rank of the mutineers, and as we all know, Nature took its inevitable course. Edward learnt a lot from this experience, and would eventually go on to become one of the most successful pirate captains of all time, in addition to pioneering radical new designs in the haircare industry. In fact, let's put Eddie's success in scope – he ended up getting a full pardon from the governor and marrying a wealthy landowner's daughter. Imagine being literally the most feared criminal to terrorize three empires and then getting one of those empires to give you a full pardon while you cheerfully walked off to shack up with North Carolina's equivalent of royalty. To translate that into a more timely historical narrative, that's like Bonnie and Clyde getting a full pardon from FDR, quitting their life of crime, and going off to become rich Hollywood celebrities. The odds of something like that happening are lower than the odds of rolling snake eyes in a game of craps, every single time. But Blackbeard pulled it off somehow. Near the end of his career, we could easily describe him as the quintessential "man who had it all."

Who can say why he decided to throw it all away when he had everything he wanted? Perhaps it was the allure of the sea, pulling him back to it in a way that only a pirate could understand. Either way, Edward made a fatal mistake – he dove back into a life of crime for "one last job" when he already had everything he ever wanted, and the crazy and unpredictable nature of the sea finally did him in.

In folklore, the story doesn't end there. Some say that the Angel of Rationality had compassion for Teach, and let his ghost stay in this world after he died, wandering the seas that he had always loved. But this is a parable about rationality, not superstition, and so we should focus our attention on the central lesson of this tale rather than diving too deeply into folklore.

The lesson of the parable is this: once the game has been changed so that a particular behavior, philosophy, or code of ethics is no longer game-optimal, people's behavior will change very rapidly to whatever behavior best suits the new static equilibrium point, and they will rationalize said change in whatever way is most convenient. This is how ideas spread, because at the end of the day, an idea is nothing more than a tool to upgrade one's *historical narrative*. Since more modern *historical narratives* flourish while outdated *historical narratives* die out, ideas that have more utility to Life end up reproducing naturally.

The point of all this is that if you can make an idea which can both spread predictably (as Tesla's ideas did) while replicating fast (as Teach's ideas did), then you have effectively created Life. Congratulations! Now let us discuss how you can get said Life to work for you.

THE POWER OF WHISPERS

In the foreword of this book, we discussed how a hypothetical ancient Aztec leader brought to the present day would doubtless be awed and amazed by some of the wonders of modern technology, such as the Internet, robotics, and avocado toast. But would we be equally as impressed with the wonders of his era? What if he were to tell us that he could use the magic of his people to whisper a phrase into an empty room and change the course of history? For example, to send a crucial message to our President or another foreign leader, or perhaps to an influential performer, scholar, or priest – all while letting the universe itself shape the trajectory of the message until it reached its recipient? Doubtless we would laugh at the idea of magic existing. But in theory, the Dark Arts of Rationality allow us to do exactly what he described here, and these Arts are not dependent upon any sort of modern technology – only the ability to analyze and understand group behavior. Game Theory allows us to craft a message that will replicate predictably, and Memetics ensures that the message spreads until the data packet strikes its desired destination. So maybe when that hypothetical Aztec told you about his priesthood's "dark arts of magic," he was actually talking about the Dark Arts of Rationality.

All this dry theory is getting boring, so I should give you an example of how a proper *whisper* works. Or perhaps it would be better to say a controlled demonstration, since it is impossible to **say** something and then **unsay** it. Once a piece of information is out there, there is no getting that genie back into the bottle. So, you know… I guess it's always good advice to tell you that in these days of social media, it's always important to be careful what you **say**.

The basic idea behind a *whisper* is that you are carving out a new trench into the incentive topography – ideally a very narrow and deep trench. Since people's choices flow along the path of least resistance, such a trench will inevitably divert some people into making choices that align with the topography of those incentives.

Phrased in such an abstract way, this may not make very much sense. Allow me to give you an example. Let's analyze the incentives of the terrorist group known as Antifa. Antifa's primary purpose appears to be terrorizing people whom they consider "Nazis," an ever-expanding definition which ranges from "KKK members" to "Trump supporters" to "anybody who is a beneficiary of any sort of 'privilege'." Because the politicians in the states where Antifa runs rampant benefit politically from this terrorist organization's support (both in terms of getting members votes and intimidating their opponent's voters), they turn a blind eye to the violent terrorist behavior of this group. Due to the *Virtue Signaling Escalation* principle, this situation predictably creates a feedback loop that encourages Antifa to become increasingly more violent in their attacks, since they gain an incentive (social status) through their attacks against the rival tribe, without a corresponding disincentive (legal punishment) to deter them from these attacks. Another memetic custom that Antifa have adopted is the practice of wearing masks while engaging in their terrorist activities, since the ability to hide their identities during their violent illegal assaults makes it easier for them to avoid punishment. Politicians in the states where Antifa roams could easily solve this problem by criminalizing the wearing of masks during protests. With the advent of mass self-surveillance due to the propagation of smartphones, this one simple change in the law would reduce Antifa violence significantly since politicians and the state officials that they appointed would no longer have any face-saving excuse to avoid prosecuting Antifa crimes. Of course, since the politicians of those states benefit politically from Antifa violence, they have absolutely no incentive to reduce said violence, which is why the "anti-mask" laws that many other countries have adopted somehow never seem to get passed in states like California. For politicians in those states, the violence against Republican supporters is a feature, not a bug.

So now that we've analyzed the ideological structure of the Antifa egregore, how can we use a *whisper* to domesticate this wild critter? For starters, remember that information changes people's motivations. Changing what people **know** also changes what they **want**. We convey information through a process called language. Language creates a data packet (also known to laymen as an idea) that our minds process and then filter through our incentive structure topography to choose the course of action that is deemed most beneficial to Life – in other words, **us**. We then rationalize this course of action in whatever way makes us feel like we made a conscious decision.

This means that all we require in order to change what people want is to change the data packets of information that they are exposed to. We just have to *whisper* the right words, releasing an idea designed to change hearts and minds. By incorporating *Tesla's principle of weaponization* (to ensure that the idea is assimilated) and *Teach's principle of preference cascades* (to ensure that the idea spreads) we can ensure that the message eventually makes its way to the person (or people) whom it is supposed to target. Effectively, we have "made the idea alive," turning it into an egregore.

Now some rationalists might question the *usefulness* of this. They might say "Big whup. So what? I suppose you could theoretically get a message to literally anybody in the entire world – a president, a celebrity, a venture capitalist, a head of state, or even a religious figure – simply by whispering the right words to some rando on the street, or making an anonymous post on 4chan. But how could a talent that obscure possibly be *useful*?" I hope we can all agree that these people's vision is a bit limited in scope. Personally, I think that the ability to pass an idea to literally anybody in the world that gets them to change their mind about a particular course of action is tremendously *useful*, but perhaps this is simply an innate philosophical difference that distinguishes a scientist from an engineer, or a Rationalist from a Dark Rationalist. Where the first group comes up with abstract ideas that have little inherent use, the second group converts those ideas into design principles that have real-world applications.

Anyway, back to our example of how to craft a *whisper*. In this specific case, we want the data packet that we are releasing into the world to be a piece of information that will modify Antifa's behavior by changing the incentives and disincentives available to them. To ensure that the information spreads to where it needs to go, we must ensure that the data we are providing meets two specific requirements:

1) It must be something that can be used as a weapon (*Tesla's principle of weaponization*)

2) People must be incentivized to spread the idea (*Teach's principle of preference cascades*)

What idea could we craft that fits that description? Well, for starters, we know that one of Antifa's memetic strengths is the fact that they wear masks all the time, thus allowing them to commit violent acts without legal prosecution. We could point out that this memetic strength is also a memetic weakness under the right circumstances. For example, in any large group of masked people, it is trivially easy for any infiltrator wearing a mask to slip in carrying an explosive device.

We could further point out that while California politicians are currently incentivized to turn a blind eye to the depredations of Antifa and their terrorist behavior, it is very hard for them to continue this policy of informal permissiveness if explosions happen as a direct result of these people's protests, because voters tend to frown upon explosions in their electoral district.

We could additionally make note of the fact that it would be trivially easy for any such saboteur to frame Antifa for the act simply by calling in a bomb threat from a burner phone immediately before executing the attack and reporting that they saw a member of Antifa with a leaky or dangerously volatile-looking bomb. This means that the explosion would very likely be attributed to a homemade explosive device that had gone off prematurely.

Do you see how this is all very dangerous information? That is the nature of a *whisper*. The information **needs** to be dangerous and valuable in order to replicate. If it doesn't possess that characteristic, it doesn't have the qualities of Life and therefore will fail to spread. Anyway, I feel this is a perfectly adequate data packet to change the existing incentive structure, so let's call it a day and say that we've created a reasonably competent *whisper*.

Now that we have crafted the *whisper*, we now need to release it out into the world. This is done simply by sharing the information with somebody else. For example, you have already helped the *whisper* spread simply by reading these words. The idea is in your minds now. You know how trivially easy it would be to destroy Antifa simply by exploiting a flaw in their memetic values: in other words, this idea can be used as a weapon. And you will eventually tell somebody, because when you are trying to explain the subtleties of how one can reshape group behavior simply by releasing a data packet out into the world, everybody is going to look at you like a retarded puppy when you explain the abstract principle without giving a concrete example of how it works. People are just too unfamiliar with memetics to understand how it works without an example of how it might look in practice, because memetics is so far out of most people's day-to-day experience that their brains simply cannot comprehend it. So even if you are positively disposed towards Antifa, you will spread the data packet because the ability to craft *whispers* is a *useful* weapon and you want your side to have it, which means that you need to explain how it works… which in turn means that you need to give specific examples.

This means that the data packet will replicate until it hits two specific targets:

1) Radical right-wing extremists who are heavily incentivized to use the tactic that I just described against Antifa,

2) Antifa themselves.

In other words, you have just given Antifa's enemies the perfect weapon to use against them as long as Antifa continues to wear masks. Meanwhile, you have also made it clear to Antifa that this weapon is now in their rival tribe's hands. At this point, they face a difficult choice: either they can continue to wear masks to avoid the legal ramifications of their crimes (in which case it is only a matter of time before somebody starts using this tactic against them) or they will stop wearing masks to protests (in which case they can no longer use anonymity as a shield to protect themselves against the law). Either way, Antifa violence will go down – in the first case, because of a sharp reduction in the number of Antifa members themselves, and in the second, because they will now be fully accountable for their own violent actions due to the fact that their faces are being captured on video during protests. Congratulations! You have just changed group behavior.

This leads us to an interesting philosophical question. Suppose that something really destructive eventually happens as a result of this *whisper*. Based on our personal framework of ethics, how can we possibly apply blame for the outcomes? In the classic sense, who is to blame for this?

Some might say that the far-right extremists committing the actual crime are to blame for this, since it is their actions that are directly causing the violence.

Others might say that you readers are to blame for this, since you are spreading dangerous information which you know will eventually lead to this kind of attack. I didn't **make** you share this information: you are doing so of your own free will.

Still others might say that I am to blame for this, since I came up with the original idea, which it's doubtful you would have had on your own.

More refined legal analysts might say that I wouldn't have chosen this specific idea to be my example if I hadn't personally been threatened by Antifa members in the past simply for making arguments in support of Trump, which they incorrectly interpreted as me being a "fascist Nazi." Furthermore, I'm not forcing these Antifa members to wear masks while breaking the law: they are knowingly **choosing** to do so despite the risks while simultaneously engaging in criminal activity. Play stupid games, win stupid prizes.

People who are starting to make the shift to consequentialist ethics might say that Antifa's leadership is to blame for this, since I wouldn't be feeling so vindictive against their entire organization if I myself hadn't been radicalized by the *expanding circle of retribution* principle. In other words, if the Antifa organization had a governance system which allowed me to successfully petition for punishment against the two **specific** Antifa members who threatened me, then I wouldn't be getting mad at their entire group and talking shit about them.

People more advanced in consequentialist ethics might say that California politicians are to blame for this, since they could easily have predicted that Antifa would become increasingly violent due to the *virtue signaling escalation* process and crafted legislation that would mitigate this group's violent behavior.

When those politicians attempt to defend their policies by pointing out that they didn't know about any of these principles of group behavior, I might reply that I have known about many of these principles for two decades and tried to explain them to people, but because these principles are all derived from evolutionary psychology (a branch of science that many of today's "woke" idiots consider racist and illegitimate), I had to be very circumspect about what I said for fear of being accused of racism by the online outrage mob and getting fired.

I might also mention that I have tried to provide advice to Antifa in the past to structure the governance of their organization in ways that would have made them operate more effectively so that they would not be vulnerable to tactics like this, but my opinions were shot down due to the fact that as a white cis male, my ideas were considered less valuable than those of historically underprivileged minorities. In other words, I was told that I should check my privilege and stop "mansplaining."

In regards to blame, I suppose the only thing that can be conclusively said is this: when somebody skilled at Game Theory offers you *useful* advice about how to structure your organization more effectively to avoid systemic problems, perhaps it is a good idea to simply say "thank you" and start implementing their proposals.

THE ANCIENT AND HONORABLE HISTORY OF MEMES

In the last section, we discussed how to create an idea that mimics the characteristics of life: in other words, what I would define as an egregore. If this seems really complicated and difficult, that's because it is. Creating and raising life is never easy – just ask any mother. Creating a brand-new idea that is more efficient and rewarding than anything currently in use requires a very high level of intelligence. That's the bad news. The good news is that we don't need to create these new ideas ourselves: our ancestors have already done most of the hard work for us. There are several really powerful ideas already floating around in a dormant state. All we need to do is rediscover one of these ancient ideas and retrofit it with fresh memes more suited to the present day, and it'll spring right back into life like a mushroom spore dropped into a viable environment. The funny thing is that most people have such a poor grounding in history that they won't even notice that this "brilliant new idea" is actually quite ancient. In fact, I would be willing to bet that you could take one of the oldest ideas in the world, slap an Instagram filter and an upvote button on it, and most people wouldn't even suspect that this trendy new thing that they just discovered is actually an ancient religion that has been around for millennia. Needless to say, this saves you a lot of time and effort on R&D. Also, ancient ideas tend to be the most powerful anyway – that is why they have survived so long. Entropy is a very powerful force so any egregore that can resist it for millennia is obviously something with a corresponding amount of staying power.

It may be *useful* to clarify what I mean when I talk about "retrofitting an ancient idea with new memes." I already gave you one example at the end of chapter 2, where I discussed some of the failure modes of the Catholic Church and how it could be improved. The Catholic Church is one of the wealthiest institutions in the world, yet user engagement has been steadily dropping. That is because the Church is, for the most part, still living in the past – using techniques and methodologies that are outdated. Putting a "Black Lives Matter" sign outside of a church (while completely neglecting the Church's digital presence) isn't real structural change: it is simply the **appearance** of change. In all fairness, many politicians, entrepreneurs, and institutions make exactly the same kind of mistake. A lot of people's default behavior is to slavishly obey the status quo while simultaneously presenting themselves as brilliant innovators. This is best described as "a strategy that works really well until it doesn't."

For an idea to spread rapidly, it must be a simple concept that is *useful* to Life. This encourages Life to spread the idea, because individuals that adopt the idea into their mental toolbox tend to be more successful than individuals who do not. Over a long period of time, the process of natural selection leads to the people who utilize good ideas flourishing and the people who reject good ideas dying out.

We each have our own cultural biases and are individually programmed by our societies and class structures, as well as a wide variety of other factors – so it can often be difficult for us to be objective enough to understand which ideas are *useful* to life and which are not. One good rule of thumb is this: any idea which has lasted for over a thousand years and recurred multiple times throughout different societies probably serves a pretty useful evolutionary purpose and deserves closer examination.

Because ideas frequently go dormant and then reincarnate in a more updated form, it can be hard to recognize them as the same idea in a slightly different format. It is easy to recognize how the pantheon of Greek gods such as Aphrodite, Ares, or Hermes ended up morphing into Roman gods like Venus, Mars, or Mercury. Unfortunately, most recurring ideas do not translate into other cultures in such a straightforward manner. To do so, we need to keep a sharp eye on recurring themes. Often the same symbols may pop up again and again throughout different cultures. Heraldry, fashion, and religion often demonstrate recurring themes, and studious historians often spot similarities which are attributed to "cultural drift." I do not dispute that cultural drift happens, but one area of study that I feel most historians neglect are the **reasons** behind cultural drift. Ideas do not spread randomly; they spread because they are *useful* somehow, and if we can identify the *useful* parts of those ideas then we may find valuable tools.

You've already read a few historical parables of mine. The explicit purpose of a parable is to teach a lesson. I like to make my parables historical because I feel that there are a lot of instructional moral lessons we can derive from studying history. For example, what could we say is the moral of the story in the parable of Marcus the Plebian? Personally, I would say that the moral of the story is that we should not subsidize systems or people whose existence is detrimental to our own well-being, because that's just stupid and it leads to bad outcomes for us. You might say that this is not such a deep fundamental insight, but I see leaders of Westernized nations making this mistake **all the**

time: giving foreign aid to countries whose continued existence does not seem to be in any of our collective best interests. There are all sorts of motives that our leaders spin up to rationalize their own actions – compassion for the less fortunate, guilt about colonialism, a cowardly lack of spine – but at the end of the day, who cares about all of that? Due to our global leaders' inability to fix climate change, we are shortly about to enter a period when resource scarcity is going to be a real problem and based on all of the available scientific data, I really don't think that everybody on the planet is going to survive. So when you think about it that way, I'd rather see most of my country's tax dollars working to ensure the survival of collaborative and pro-social societies rather than intolerant foreigners who seem to have very backwards values in regards to issues like free speech, gender equality, LGBT issues, and social safety nets. I don't particularly care about the ancient history of whether our ancestors oppressed their ancestors or their ancestors oppressed our ancestors before that. All I care about is that American kids are going to have a lower chance of surviving the coming climate apocalypse, specifically because our politicians seem to feel that subsidizing the continued existence of backwards hellholes where women sometimes get raped on a daily basis and gay people get stoned to death is somehow more important than ensuring the continued existence of more civilized nations where these kinds of abominations are strongly discouraged. And I think if you asked most other people to think about that really carefully, highlighting all the undeniable data which points to the fact that the end of the world has started and not everybody on the planet is going to survive this, they would have a similar reaction. I think that once you got all those people through the various phases of grief, they would eventually wipe away their tears and say "Look, I want our future to be a utopia rather than a dystopia. And if the sad realities of climate change mean that some people are going to have to be sacrificed for the greater good – then as terrible as that may be, I want it to be the **bad** people of the world who get sacrificed. We should at least initiate a serious conversation about whom the bad people are in this scenario, and then politely but insistently request that our politicians start spending tax money in a way that maximizes **our** collective well-being rather than the survival odds of potential future adversaries." And here's the important thing – being realistic about the grim odds of the situation that we're all in doesn't make you a bad person. The terrible people of this world could make a choice at any point in time to stop their bad behavior and start behaving in a cooperative way that will help all of us advance together. They **choose** not to do this; instead they make the **choice** to persist in their backwards behavior and blame us for their own moral failings. Now I'm not suggesting anything wild and crazy like sending AI-controlled robot death squads out to foreign countries, but I do want to point out that we are to a certain extent **teaching** those countries how to treat us through our own reactions to their behavior. If we respond to bad behavior by rewarding it, then we are to a certain extent encouraging other people to treat us poorly. That's just not good teaching technique.

You see? The parable has such a simple and self-evident lesson, but when you unpack it, it really raises a lot of questions about the state of the world around us, as well as the competence of our leaders in general. If they're somehow lacking basic common-sense life lessons like "Hey, we shouldn't reward bad behavior – we should reward good behavior instead" then it raises questions about their stability in other matters.

But if we're going to have a conversation about how parables are nothing more than simple and self-evident lessons, then perhaps we should let this next one be taught by a narrator with a bit more experience in the art of instruction.

The Parable of Aleister Crowley

"We have a problem, Edward," Yeats said, slamming the door closed behind him so forcefully that the tower of tarot cards collapsed to the desk. Aleister sighed, and began collecting the toppled cards to reshuffle back into the deck. "Do we, William? That is shocking news."

"Your sarcasm towards a superior member of the Order is duly noted," Yeats said. "Just another attitude problem to add to a very long list."

"And what list would that be?"

"Don't play the fool with me please. Out of your numerous moral failings, it's strange that you would choose to display the only fault that you do not in fact possess."

"You sound very gifted at discussing moral failings, William," Aleister said. "Perhaps I simply wish to gain a more robust perspective of such things... from a superior member of the Order, of course."

"Fine," Yeats sat down brusquely. "Let me spell it out **simply** for you. We are a society of scholars and visionaries. Magic is a sacred art that needs to be spread to the masses, so that they may find spiritual enlightenment. To spread this doctrine, we need to set an example for the common man to follow. You fornicate freely with women of loose virtue, ladies of the night, and even with other men, or so it is rumored. Do you really need me to explain why this poses some problems, Edward? Do you understand why it is difficult to spread the doctrine of magic when your unsavory practices cast such a long shadow upon its very reputation?"

"I must indeed be a fool, William, because it is my impression that I was doing the Order a favor."

"A favor?!? I – you -" Yeats stammered to a halt and clenched his fists, his face flushing with anger. "Please explain to me how in God's name you feel that this kind of licentious behavior does any of us a **favor**."

"I hesitate to educate a superior member of the Order, for fear of seeming insubordinate and -"

"Just shut up and explain yourself, Eddie."

Aleister smirked, and began reshuffling the tarot deck. "It's quite **simple**, William. We are salesmen. Magic is a pragmatic idea that is also a tool. We want to sell that idea to the masses. So obviously we need to package it in whatever format is most appealing to them."

"That's exactly what I've been trying to tell you! We need to set a good moral example for them to follow, so -"

"Do we?" Aleister asked. There was a short silence.

"Well, of course," Yeats said. "The masses want –"

"With all due respect to a 'senior member of the Order,' I don't think you have the slightest clue what the masses want," Aleister interrupted. "The Church made up an arbitrary morality for them to follow and you assume that they aspire to be devout paragons of that morality, but I don't think that's what they **really** want. I think that if freely given the choice between the restrained life of a distinguished gentleman such as yourself or my unrestrained life of power, fame, and sex, most would choose the latter."

"I don't follow," Yeats said hesitantly.

"Oh, but you do, William. You follow the Church, which teaches us that in order to be considered upright moral men we need to restrain our natural passions. That when we are wronged, we need to 'turn the other cheek' instead of vindictively striking back sevenfold. That when we identify something we want, we should strive to repress our desires instead of striving to fulfil our ambitions. And if that kind of lifestyle actually appeals to you, that's fine – I certainly won't judge. But why on earth would you consider that sales pitch to be something that the common man would find *useful*? The

Church already sells that philosophy with far greater enthusiasm and vigor than the Order does. I think that if we want the idea of magic to spread, we need to package it as something different, not just more of the same."

"This sounds like nothing more than a tissue-thin rationalization for your own hedonism and debauchery."

"What better way to sell a narrative than by practicing what I preach? The Church sells restraint, humility, and servitude. I sell sex, fame, and wealth. And I prove to the common man that those things are all attainable, simply by practicing what I preach. No empty promises here: what you see is what you get. I use magic to live well, and others aspire to be like me, emulating my techniques, and so the practice of magic spreads."

"And to hell with the rest of us who disagree with your methodology." Yeats said bitterly. "No. I won't tolerate it."

"I didn't think you would. Tolerance has never been your strong suit."

"And what is your strong suit? I have heard your mantra: 'Do what thou wilt.' I know that you are selfish, degenerate, and base, but you claim to care about the future of civilization. What possible future can it have when everybody just... does whatever they want?"

Aleister took a deep breath. "Was that a real question?"

"What?"

"Was that a real question? Or are you just trying to score points? Because I think it's a really good question – one of the few that you've asked – and it deserves a good answer. But if you are just posing the question rhetorically to score points, then I'd prefer not to waste time answering you when there are much more productive ways to spend my time... such as building another card tower for you to knock over."

"Very well. I'm listening, Edward. I don't expect to hear anything that will update my opinion, but I'm listening."

"Thank you, William. We haven't always seen eye to eye, but you do at least take the time to listen to others. Credit where credit is due." Aleister paused for a moment, thinking about how to frame his response. "If I am understanding your complaint correctly, you are telling me that my morality doesn't translate well to the universal level. Society wouldn't function if we were all to fall into the trap of unrestrained hedonism."

"That is a mild way of putting it, but yes."

"Well then, since you are such a devout Christian, consider the philosophy of Jesus. Non-violence was a core part of his philosophy. But how long would our empire last if every man in England adopted his teachings and refused to spill blood? We would be conquered by the first invader to hit our shores. Why, a child with a stick could conquer our empire if England were to adopt the teachings of Christ universally."

"Wonderful. So now, in addition to your numerous character defects, you are choosing to blaspheme against our savior."

"Not at all. Jesus was the right man for his time, just as I am the right man for mine. He lived in a time period where there was precious little compassion; where criminals were thrown to the lions for sport. At that point in time, the world needed more compassion, and Jesus Christ brought it to them. I am simply pointing out that Jesus's message of compassion would be just as damaging to society as my own message of freedom, if England were to truly adopt such a message universally. Nobody ever adopts a virtue universally, and that's a good thing. Prophets simply bring whatever virtue is most needed for their time, to fill a void that their society lacks, or to counterbalance a past virtue followed to excess. Eventually, society changes enough that a different virtue is needed, but that is a job for the next prophet."

"So now you aspire to be a messiah figure? This just gets better and better." Yeats shook his head in disbelief. "And just what is it our era lacks that you claim to bring to the table?"

"Tolerance," Aleister said calmly.

"There are things that the world should not tolerate."

"Such as intolerant people?"

Yeats took a deep breath. "Aleister, please try to understand this... you are delusional. I know that you think you are on the right side of history, but this world that you think you are instrumental in building will never come to pass. Do you seriously think that the masses will ever accept women of loose virtue into the upper echelon of society? That they will care about the well-being of prostitutes, as if they were normal human beings? Or

– most laughable still – that it will ever be accepted for men to lie with other men? This vision you have is nothing more than a fever dream, the megalomaniacal delusions of a drug-addled maniac. You are stark raving mad, Aleister, and this 'guardian angel' whose dictates you claim to follow is nothing more than an echo of your own dark soul." He turned and left, slamming the door behind him. Several tarot cards blew off the table and onto the floor. Aleister picked one up and examined it carefully, then slid it back into the deck.

"We shall see." Aleister whispered to the empty room.

And eventually, they did.

What is the intended point of this parable? Obviously, everybody will infer a different meaning from it, but the point that I **intended** to make is that what society considered "righteous" behavior in one era – treating sex workers like trash, demonizing homosexuality, or expecting women to live up to unreasonable standards of virtue – would today be considered the most backwards savagery. Literally every time period of history has some behavior that looks horrific in retrospect. So why should our own time period be any different? Frankly, I don't think it is. I think that a lot of the behavior that our society currently views as righteous upstanding behavior – such as legally discriminating against people of the "wrong" race, claiming special status or superior insight due to one's sexual orientation, punishing people of a particular gender or race more harshly for crimes, or silencing those who speak up against such treatment – will in the future come to be viewed as behavior equally as evil as that of the society that Aleister Crowley lived in. Similarly, I believe that the descendants of the people who supported this disgusting bigotry and attempted to rationalize excuses for it will be embarrassed about their connection to these hateful figures who rightfully belong in the trash-heap of history. So why not just cut to the endgame, accelerate the process, and expose them for what they really are?

To be even more specific, our leaders often talk about having the "political capital" to solve problems. This policy can't be implemented, no matter how positive, because voters would never support it. Or that policy can't be given support, because it would be perceived as too harsh. My point is that properly deployed memetics can literally change what the voters want, so obviously most of our leaders don't have the slightest clue what they're talking about. We elect them to solve problems and they stumble around cluelessly, claiming that their actions are totally constrained by us, the voters. But in reality these constraints only exist in their own minds and they already have all the tools that they need to build the consensus necessary to change society. While we may as individuals have strong opinions, society as a whole doesn't know what it wants, and it is willing to be led by anybody who has a plan that is obviously beneficial for them. If you have a solid enough memetics campaign, the public will want whatever you **tell** them to want. In other words, we have all the tools we need to build utopia already within our grasp – the one thing we lack are pragmatic leaders who are willing to efficiently deploy these tools to maximize results. My hope is that this book inspires some of you to be those leaders. No matter how average you may think you are, simply being able to apply the tools of Dark Rationality towards a problem makes you far more qualified for leadership than 99% of the so-called "leaders" and "elites" currently holding those positions in our society. Don't be afraid to use these principles to reach out for what you want in life. The only thing truly standing in your way is **you**.

Chapter 4
~~Prayers~~
~~Spells~~
~~Instruments~~
~~Weapons~~
~~Tools~~
Techniques

BUILDING YOUR TOOLBOX

So far, all we have discussed are the abstract principles of rationality, but very little of the applied sciences. If you have managed to read up to the point, you may understandably be getting a little impatient. Considering that this book is named "The **Dark** Arts of Rationality" (indicating a more pragmatic focus), it would be a bit of a letdown if I focused exclusively on theory and neglected the application.

The purpose of this chapter is to give you a few ideas about how all of the abstract principles discussed earlier in this book can be leveraged in various ways to get you what you want out of life, whether that is money, fame, political power, or simply the ability to improve the world more effectively. The following techniques are not a fully comprehensive list – since this book is a simple introductory primer into the Dark Arts of Rationality, I have stripped these pages of anything that might be too dangerous in the hands of the public. We will thus focus only on the least hazardous implementations of the Dark Arts, such as making money, improving or destroying large organizations, brainwashing people, and learning secrets about others that they might prefer to keep hidden even from themselves.

For purposes of efficiency, each technique is categorized by name and school. A basic explanation of the principles behind each technique is given followed by a detailed description of how those principles can be leveraged to achieve your goals.

MEMETIC HAMMER (Memetics)

How it works

Memetic hammer is a quick and dirty way to hammer a low-level concept (such as anger, compassion, sex appeal, religious devotion, etc) into somebody's mind. It is *useful* as a reinforcement device for other memetic techniques, as long as it is used subtly and strategically. It works by exploiting a cognitive defect in human pattern recognition – namely, the fact that exposure to an idea creates a neurological path in our minds that is automatically reinforced through repetitive exposure. In other words, we tend to favor patterns that are familiar to us, regardless of whether those patterns are good for us or not. Memetic hammer exploits this by rapid repetition, quickly hammering mental patterns into our minds without grace or finesse – but with a lot of effectiveness. Part of the effectiveness comes from the fact that human beings have adapted to pick up micro-expressions from other people, which causes subconscious responses in us. By displaying those micro-expressions over and over again, that same primitive feedback loop is triggered repeatedly.

How you can exploit this

Plant articles in viral infotainment aggregators that have repetitive gif clips tailored to evoke a specific emotion in the target. Thanks to the fact that we are attracted to familiarity, each repetition of the clip embeds the concept a bit deeper into the target's mind.

For example, if you are a politician trying to make yourself seem more relatable to the public, propagate a repetitive clip of yourself doing something that the public considers appealing. A clip of a tender gesture towards a spouse could make you seem warmer and more empathetic, while a dominant gesture could make you seem tougher. It all depends on the image that you want to convey. Likewise, saturating the airwaves with repetitive clips of your opponent's least photogenic moments is a good way to make them seem dopey, evil, or out of touch.

Obviously, politicians are not the only ones who can benefit from memetic hammer. Celebrities benefit from selling sex appeal, CEOs benefit from selling the appearance of intelligence, and salespeople benefit from selling the usefulness of their product. Meanwhile, those who are in the business of collapsing societies benefit by highlighting corruption and rampant inequality, or the appearance of such. Websites such as "Rich Kids of Instagram" are often very helpful in that regard, since young people are often too stupid and self-absorbed to realize when they themselves are laying the foundation for revolt against their elders.

One important caveat to remember is that memetic hammer is a primitive methodology which is not very effective on its own. I would define it as a tool to lay the foundation for a more sophisticated memetic attack rather than being an attack itself, since it is meant to prime your target audience with subconscious expectations, making their minds more susceptible to the message which you wish to deliver. From a defensive perspective, if you are very observant, you can often anticipate an opponent's strategy by being alert to the social priming cues that you see them deploying beforehand through the use of memetic hammer.

ANGLERFISH AND SUBLIMINAL MESSAGING (Memetics)

How it works

Subliminal messaging is a way to manipulate an audience without triggering their psychic defenses, allowing you to slip messages into their mind "under the radar" without them being aware of it. Used by an expert, it can be used to do things like start political movements, generate romantic attraction, or (under our current capitalist model) get people to buy lots of useless crap that they don't really need. Anglerfish is the visual version of this (deployed to great effect by certain entertainment celebrities) but there are multiple modalities that can be exploited.

How you can exploit this

Due to the bizarre environments and clothing needed to make subliminal messaging effective, it is primarily exploitable by musicians because the flashy environment of a concert synergizes well with these techniques, allowing them to be deployed without anybody noticing.

For example, due to the *principle of pattern-matching*, we know that what you focus your attention on generally becomes more attractive to you over time. In other words, if you simply look at a person for long enough, that person becomes more attractive to you. This is why successful musical celebrities often embed easter eggs in their music videos for fans to find. When fans view the videos multiple times in order to find all the clues, they are also exposed to images of the celebrity, which makes the celebrity more attractive to them, which makes them watch the videos more, etc. Feedback loops like this are very *useful* in terms of manipulating the emotions and behavior of audiences.

What makes concerts the perfect venue to deploy this technique is the fact that lighting and clothing can be used to amplify the effectiveness of this attention grab. People have known for a long time that clothing can be used to accentuate or alter various physical features to instill a psychological effect: for example, making people look more attractive or intimidating. What they don't know is that modern technology can be used to trigger these psychological stimuli much more strongly. For example, our attention is designed to focus on motion, because in the primitive environments that we evolved in, motion was indicative of either potential predators or potential food. This can be exploited by smart clothing to subtly but repeatedly draw our attention. An entertainer using this technique might make an artistic statement by wearing color-changing smart clothes with an animated snake slithering around her body. Audiences won't be perceptive enough to notice that the snake spends a statistically more significant amount of time around sexualized body parts, such as her breasts and legs, because normal people don't spend much time introspecting about their own feelings and correlating them to external data. The only thing that the audience might notice is that they mysteriously find themselves more attracted to the entertainer, like magic. But the only magic at play here is psychology – **real** psychology, not the politicized feel-good garbage currently taught in our universities.

Lighting is also an effective tool for this purpose because our minds have a limited bandwidth, and the more external stimuli that they are processing, the harder it is to identify and resist subliminal messaging. Using light patterns that shift subtly (enough to disorient but not violently enough to trigger a fight-or-flight effect) makes people's minds less able to resist subliminal messaging. This means that any skillful performer using these techniques can slip a surprisingly large amount of brainwashing in under the radar to make the audience feel good and keep them coming back for more. And isn't that what they're paying you for?

INCENTIVE UNBALANCING (Memetics)

How it works

Incentive unbalancing is a long-term strategy to destroy large organizations by causing them to self-destruct. This is accomplished by slipping structural defects into your opponent's organization which create dysfunctional resonance effects that amplify each other with each perturbation. Due to the fact that these structural defects amplify and build off each other, even a small nudge in the right direction can sink giant organizations, given enough time. The entertaining part is that once the dysfunctional elements of the organization are firmly in place, over time stakeholders who exploit this broken system will develop a personal investment in the organizational dysfunction and will actively resist any attempts to solve the problem because they benefit from it. From a defensive perspective, the best way to defend against an Incentive Unbalancing attack is through a robust program of Incentive Alignment (described later in this chapter). This is definitely a situation where an ounce of prevention is worth a pound of cure.

How you can exploit this

At the current moment in time, the best way to take down an organization is to stack its HR department with hard left ideologues. This process, commonly known as "Roll hard left and die," has a lot in common with the *Virtue Signalling Escalation* loop. When a corporation starts failing or has a series of setbacks, it is in the long-term best interest of everybody who works there to address the problem in a realistic way, seeking out constructive criticism to improve their own business practices. However, remember that the *Defector's Dilemma* states that for any variety of problem where Life can gain a significant short-term advantage in preserving or replicating its own pattern (typically by acquiring prestige, wealth or power) by betraying its own principles, it will almost always do so, even if this damages its own long-term position. This means that seeking out constructive criticism is the absolute last thing that most high-level executives are incentivized to do. While it may be good for the corporation as a whole to acknowledge when executives make strategic errors, it tends to be bad for those executives' careers. Therefore, it is in their interest to attribute blame for their failure on outside factors. In our current era of woke lynch mobs, the easiest outside factor to blame for corporate failure is racism, sexism, or any other sin of insufficient progressivism. In our modern environment of outrage mobs and microaggressions, racism and sexism are such ambiguous and poorly-defined terms that any legitimate criticism of the business plan can be spun into a politically insensitive move. This means that such excuses can be used not only to hide executive incompetence but also to cast blame at any critics who wish to point out said incompetence, thus silencing critics. This is especially true when HR departments gets stacked with hard-left ideologues, because it is a known truism that witch-hunters always find witches. When an HR department gets stacked with hard left ideologues, then it becomes increasingly easy for executive leadership to silence critics with accusations of "insufficient wokeness," and therefore it becomes increasingly tempting to do so, because large groups of people tend to move in the path of least resistance. Incentive topography thus makes it inevitable that these organizations will become increasingly dysfunctional over time as the executive leadership avoids dealing with their own problems and instead purges valuable members of the company in politically motivated witch hunts.

Since it can be challenging to systematically stack an enemy company's HR department with hard-left ideologues, it is easier to gradually radicalize them by friending them with fake social media profiles and gradually immersing them in radical left-wing ideology, thus creating an echo chamber similar to a low-level Mandela effect (described later in this chapter). Surrounded by people who spout left-wing ideology all the time, your target will get the false perception that these types of irrational beliefs are more common than they are, and will shift their own belief systems to compensate for it.

If taking the time to radicalize an enemy corporation's HR department sounds like too much work, you can still benefit from this technique by using it to identify corporations that are in the process of "Rolling Hard Left and Dying" and making sure not to invest in them. For example, Google, Bioware, Twitter, and Wizards of the Coast are all organizations currently in the middle of this radicalization process, where they focus more upon virtue-signaling corporate wokeness than on cultivating and retaining talent. This means that they are all bad investment opportunities. Knowing which companies are definitely bad investments due to this basic principle of group dynamics makes it easier to maintain a good investment portfolio simply through the process of elimination.

ELECTION FIXING (Memetics)

How it works

By creating a cascading resonance effect between the *Virtue Signalling Escalation* and the *Expanding Circle of Retribution*, you create a negative process within your opponent's tribe that effectively radicalizes their people. The worst and most objectionable radicals are the ones that inevitably occupy the most time in the public spotlight, drawing widespread public hatred towards the opponent's tribe. People will then elect politicians who promise to hurt that group of people.

How you can exploit this

Embed propaganda agents within your opponent's ranks and task them with beginning the *Virtue Signalling Escalation* process, using operant conditioning to increase the rate of catalysis. This is perfectly legal and almost entirely invisible because displaying manufactured outrage about things is celebrated - some might even say actively rewarded - within the current dysfunctional structure of our society.

For example, you might have one of your agents pretend to be a Black Lives Matter activist who gradually says increasingly hateful and bigoted things about white people. Because the hostile shift in tone is very subtle and is dispersed over time, people will not notice the new agenda being slipped in subliminally – instead they will shift their own behavior to mimic the behavior of your agent provocateur. Over time, your enemy will develop an increasing number of radical extremists who say dangerously unhinged things which are viewed by your opponent as normal behavior, because your agents have deliberately normalized it. Then you can simply have your agents go dark and fade away while you turn the media spotlight on the hateful activists within your opponent's ranks (who have only become hateful because they were radicalized by your agents, hilariously enough). This allows you to trigger the *Expanding Circle of Retribution* (effectively an egregore's immune system) to create multiple destructive backlashes (to an egregore, an allergic reaction) against your opponent's group.

It is critically important to maximally engage the immune system response of the *Expanding Circle of Retribution* by drawing media attention to your opponents radicalized lunatics as often as possible, because the *Expanding Circle of Retribution* is the only part of this process that actually inflicts any real damage to your opponent's egregore. The *Virtue Signalling Escalation* creates openings in your opponent's defenses, but it is pointless to create openings if you don't exploit that advantage to deliver propaganda attacks. Not necessarily detrimental in the sense of creating potential backlash (though I must caution that this has not been thoroughly tested) but simply because it is a total waste of resources and time for your propaganda agents.

INCENTIVE ALIGNMENT (Game Theory)

How it works

Obviously with all the unconventional effects that memetics allow people to create, it is *useful* to create some sort of a defense for yourself. Incentive alignment is that defense. Most memetic techniques work by exploiting a vulnerability in the topological map of people's incentives. For example, politicians want support from local activist groups and celebrities because that translates directly to voter support and grassroots popularity. They will often respond to gestures of support from such groups with warm words, encouragement, and even the offer of a platform. This means that it is trivially easy to use the Election Fixing technique to get them to ally with toxic people who poison their reputations simply by association.

Incentive Alignment is the process of monitoring and tracking all memetic vulnerabilities to the structure of your organization, making sure that the minds of your employees/citizens/churchgoers cannot be easily hacked by outside influences. Typically, this would be a specialized role handled by a dedicated resource within the organization, analogous to a cybersecurity specialist. As a politician, you need to get somebody to monitor and track online activity of all activist groups you are linked with in order to quickly distance yourself and disavow any groups which turn out to be toxic saboteurs, or which have been radicalized by the *Virtue Signaling Escalation* process. As an investor, you want to make sure that the executive leadership of your organization has not had their internet use compromised and been trapped inside a Mandela Effect, giving them an increasingly deranged view of the world. In short, you need to accept that weaponization of Memetics and Game Theory is the new normal, and that if you do not have people skilled in the Dark Arts on your side, then you will always be a victim of those who do.

How you can exploit this

Know the tricks, and understand that with the weaponization of Game Theory and Memetics, the world we live in operates according to entirely different rules than it used to. Going forwards, leaderless organizations like Antifa or BLM are now effectively obsolete, because they can so easily be subverted. Any organization which does not have a means of imposing internal order and discipline upon its members is an organization which is already defunct and simply does not know it yet.

You cannot reflexively trust allies anymore. Whether in business, entertainment, or the corporate world, any potential alliance is also a potential avenue of attack. You will want to monitor the public actions of your allies to ensure that they are not being manipulated in a way that would endanger your cause. Where politicians once used to accept any celebrity endorsement without thinking, they must now do a cost benefit analysis since having a toxic celebrity associated with you may result in more harm than good. Essentially, we must now accept that from this point on, we live in a low-trust society. This does not mean that we cannot have allies, simply that we must now choose them more carefully and selectively. Most importantly, you need to practice mindfulness and self-awareness. In the past, it was possible for us to lie to ourselves about who we really are on the inside, allowing ourselves to rationalize self-serving justifications for why we want what we want and do the things that we do. That is no longer an option. If you do not understand yourself – if you cannot accurately map out your own incentive topography – then others can use that weakness to manipulate you. Self-deception is no longer an option. A world with less trust may seem like a terrible thing, but the silver lining is that it is also a world with less bullshit. Self-deception is prohibited for those who wish to avoid becoming the puppets of another.

What is really good news is that if your competition has not read this book, then unless they are clinically paranoid they will most likely be completely blindsided by the facts of life in this new world order, which means that you can ruthlessly exploit their naivete using the wide variety of techniques described in this book - all of which are perfectly **legal** if implemented in a smart and thoughtful manner. I encourage you to have fun with these techniques, or even indulge your creativity by creating your own. It is always gratifying to see people getting so excited by the ancient arts of logic and reason that they end up pioneering exciting new advances in the Art.

FEEDBACK LOOPS (Game Theory)

How it works

A feedback loop is a way to improve an organization's efficiency through the principle of Skin in the Game. When any sort of decision at an organization is made, it has either positive or negative consequences for that organization. If upper management ensures that those positive or negative consequences directly impact the decision-maker in a correspondingly positive or negative way, they will be maximally incentivized to align their interests with the organization. This tends to lead to both better decisions and faster response time when it comes to moving away from failing strategies.

In many ways, a feedback loop can be thought of as the nervous system of an organization. When an organism receives input through the pleasure/pain principle, that information must travel to the brain so that the organism can decide how to respond to said input. The further the signal must travel, the longer it takes to make decisions, leading to slower reflexes. Because slow reflexes are disadvantageous to survival, Life optimizes for quicker reflexes in a variety of ways. For example, octopi effectively have brains in their tentacles so that the neural signal has a shorter distance to travel. Since organizations are effectively egregores, they can optimize for quick response time in a similar way.

When a company has a failing business strategy that it needs to pivot away from, this information typically comes in the form of customer feedback and financial data. Ask yourself this: how long does it take for this data to be consolidated and presented to a decision maker who can effectively take action on it? And how long does it take for them to realize that this is even a problem at all? That length of time indicates the "reflexes" of an organization. As in nature, organizations that can respond to data more quickly have advantages over slow organizations, and this will be reflected in both their success and their stock price.

How you can exploit this

Feedback loops can be used to rapidly optimize large organizations, causing their internal processes and systems to become much more efficient. One example of this is Amazon.com and the method that it used to implement its Amazon for Business rollout.

To those who are unaware of it, Amazon for Business is a procurement system designed by Amazon to capture the majority of an organization's tail-end spend. Currently most large companies have inefficient procurement processes that require employees to submit a requisition to a buyer, who then places the order after a manager approves it. Later on, the requisitioner allocates the charge to a department, so that finance can then ensure it hits the correct budget and is taxed accordingly. Amazon's insight here was to replace this process with a portal functionally similar to their normal online marketplace, using a workflow so that the employee can place the order themselves and a manager can approve it with just the click of a button. It also has detailed reporting capabilities which can provide all the records needed for tax reconciliation.

My intent here is not to plug Amazon's capabilities. Rather, I participated in the pilot program of Amazon for Business and was surprised at how quickly the functional changes that I proposed were implemented into the product. Having been a programmer before, I can tell you that changes suggested by the customer typically take a bit of time to get adopted. My company didn't have any particular pull with a company as big as Amazon, so the speed at which they incorporated my suggestions (after I explained the utility, of course) indicated that the sales rep involved in this pilot program had a disproportionate amount of influence to cut through red tape within the company. This is important because of the principle of *Skin in the Game* – when you put decision-making authority in the hands of those who will be most positively or negatively impacted by the outcome of those decisions, it incentivizes better decision making. Most companies vest decision-making in the hands of people who are a few steps removed from the outcome of their decisions.

      The speed at which Amazon responded to my suggestions made it clear to me that they had a highly efficient feedback loop which centralized decision making towards individuals who had a personal stake in the problem. By examining the feedback loops within an organization, you can determine how successful that organization is likely to be in a competitive market. Since success is tied to a company's stock price and investment bankers have a very inaccurate picture of what traits make companies successful (due to the fact that their analysis relies on economics, a fake pseudoscience) you can exploit the market's ignorance by picking up valuable stocks at bargain basement prices.

GENTRIFICATION SCRYING (Game Theory)

As much as I would like to, I cannot take the credit for Gentrification Scrying, a technique taught to me by a cunning woman who prefers to remain anonymous. This technique is particularly useful to people involved in real estate development.

How it works

Gentrification scrying is based off a principle similar to betting markets, which provide surprisingly accurate forecasts of political and economic events due to their decentralized nature combined with the principle of "skin in the game." Essentially, it uses a different type of crowdsourced data to determine which neighborhoods are "up and coming" and thus likely to have rapid increases in property value.

It should come as no surprise that people whose livelihood depends on driving others from place to place tend over time to cluster in areas which are both safer (to reduce their likelihood of being robbed) as well as higher-traffic (to maximize profitability). What may come as a surprise is that these areas map remarkably well to the results of studies done on areas of "gentrification risk" – in other words, places where the value of property is about to rise very quickly, forcing lower-income families out.

How you can exploit this

While the societal implications of gentrification have been discussed at length by many other people (and are beyond the scope of this book), the important takeaway is that you can use this technique to map out areas where property values are likely to increase faster than normal. By walking around town with your cell phone out and repeatedly using an app to summon an Uber or Lyft driver (but cancelling your summons before the process is complete), you can see how long it would take for a driver to arrive at your destination. Neighborhoods that have higher property values will have shorter wait times, as will neighborhoods that are starting to become more popular. By doing this repeatedly, you can create a "heat map" for gentrification, thus allowing real-estate speculators to invest more accurately. One important caveat is that when comparing wait times, it is important to make your heat map at the same time each day regardless of which neighborhood you are in, since wait times during rush hour are obviously different from wait times at 2 AM. Obviously, in order for your heat map to be an accurate representation of reality, you need to control for external factors as much as possible.

Silicon Valley has a propensity for coming up with useful tools and then implementing them backwards and upside down. That said, it is only a matter of time before Uber and Lyft eventually realize the value of the data that they have access to and begin monetizing it with nifty charts and infographics. In the meanwhile, however, the only way to develop such a heat map is the old-fashioned way. The silver lining to this tedious approach is that as long as this knowledge is not yet widespread, people who understand this technique will have a significant advantage over their competition.

THE DEEP DIVE (Game Theory)

Since this book is intended as an introductory primer for people who are just beginning to dabble with the Dark Arts of Rationality, all of the techniques described above are fairly basic. The Deep Dive is an exception to this rule, because it uses a hybrid multiple-step approach towards divining the details of a person's character. I would consider the Deep Dive to be an *intermediate* level technique. The intent of including it in this book is to give you a taste of what kind of techniques may eventually become available to our society through academic research if we abandon our faulty *historical narrative* - the hubris of thinking that our scientists have already discovered all of the world's mysteries - and instead begin treating game theory and memetics as legitimate applied sciences which can be improved through experimentation and testing. On a personal note, I would like to give due credit to my fellow practitioners by pointing out that this cunning technique was not invented by me but rather by the same woman who developed Gentrification Scrying.

How it works

There is a remarkable amount of information about people's personal lives floating around on the internet. Most people don't realize how much personal information they are giving away because from their perspective, the information drips out slowly over time, and it is only by putting all of the pieces together and building a comprehensive timeline that sensitive information may emerge organically. To demystify the process, this is analogous to stalking a crush online, but the Deep Dive takes this information-gathering process to the next level by applying a rigorous scientific methodology to the process. Breakups, new jobs, trips, spending habits, romantic interactions – all of these things can be observed or inferred from things like public records, online resumes, Facebook, and Instagram. Hashtag counting can simultaneously be used to build up a thorough psychological profile, which assists in calculating your target's incentive topography.

The three pillars of profiling are: public records (Registry of Deeds, open-access government records, etc), the digital footprint (online resumes, social media, etc), and personal interaction. You need to identify significant markers at each level, record the dates, and then build a timeline.

The Registry of Deeds is a very useful tool for establishing a target's financial profile. Because government bureaucracies tend to be terrible at consolidating information in an efficient way, this search must often be conducted on four levels: Federal, County, Town, and Commercial. The Registry of Deeds is useful for showing spending habits. For example, if you look up a person and see that they refinance their mortgage every time the interest rate drops to take advantage of it, that builds the picture of a person with highly responsible spending habits. By contrast, if you see a person repeatedly refinancing their mortgage to purchase new cars every other year, that paints the picture of a bad decision-maker with terrible spending habits.

The digital footprint includes most social media tools such as Instagram, Facebook, and LinkedIn. It is useful for determining significant events in a person's life. When cross-referenced against a person's financial profile, it can help to build the story of **why** your target made certain choices. For example, suppose that the Registry of Deeds indicates signs that your target suddenly has a lot less money, such as selling their second car or refinancing a twenty-year mortgage and replacing it with a thirty-year mortgage. The digital footprint can help explain **why** your target has less money. Did he get divorced and now has to pay alimony? Well, Facebook typically shows relationship status and even when that information is private, it can usually be inferred through the pictures posted to a person's profile. Does your target have a spending problem? Look up their resume online, use Glassdoor to estimate their approximate salary, and then take a look at their Instagram profile to see what kind of lifestyle they are leading. Do they appear at a lot of really expensive destinations? Are they always wearing designer clothing? The answers to these questions help build the picture of whom your target is. Did they lose their job? LinkedIn will usually display the dates of employment for your target.

The interpersonal element of a Deep Dive requires the most finesse, and bears a certain resemblance to advanced interrogation techniques. Once you have assembled a detailed picture of a target's life story, talk to the target and try to get their own description of their life story (without mentioning that you did a Deep Dive on them, of course). It is especially important to take note of discrepancies where your target's narrative does not match the facts that you have researched, because what a person lies about is often more significant than what they are honest about. People lie for a wide variety of reasons. Some people lie to maintain their self-image, others lie to enhance their social prestige, and still others lie to gain romantic opportunities. If you can understand **why** a person lies and how that lie fits into their internal narrative, then you have learned something useful about their psychology which may be helpful in all of your interactions with them. Similarly, a person who is willing to be honest about difficult or disadvantageous truths has more credibility than somebody who tries to "bend the narrative" to flatter themselves.

How you can exploit this

The more advanced a technique is, the more real-world applications it has, and the Deep Dive is no exception. Governments and corporations can use it to facilitate background checks. Police profilers can use it to spot early warning signs of dangerous psychological aberrance. Lonely singletons can use it to analyze romantic prospects, fast-forwarding through the "getting to know you" phase in order to spot red flags quicker. And those who specialize in espionage can use it to spot tasty tidbits of blackmail. While the Deep Dive may be a lot of work (due to the lack of software that currently exists to facilitate the process), it can be very rewarding to those who are willing to put in the effort.

Mandela Effect (Bedazzlement)

How it works

By immersing the target and those in the target's proximity to a full-sensory translation of their internet use (which will soon become possible through the applied use of real-time algorithms like GPT-2), it is possible to make the target seem to be living in an "alternate reality" where the news is all fake, all the time. Because the target is subject to a full-scale distortion of their subjective reality, when they are released from the effect it will seem to them as if they have slipped into an alternate universe where trivial details (or sometimes very significant details) of reality have changed. This could hypothetically be used for hilarious practical jokes. To those of more boring but perhaps more practical ambitions, using this technique on rival leaders to distort their view of reality will doubtless hold more appeal. However, the usefulness is limited by the fact that the target can break out of the illusion by "disbelieving their own reality" – a challenging and problematic feat, but nevertheless one which is quite possible for somebody knowledgeable in the ways that these algorithms work. The tricky part is that you can't do anything crazy to "make your saving throw against illusions" (like stepping off a skyscraper thinking that gravity is a myth) because if the specific thing you're disbelieving **isn't** an illusion, then you have a very real problem on your hands. If you ever find yourself targeted by this effect, then the best way to escape is to rebel against the illusion in subtle ways – for example, using a whisper effect repeatedly until whomever is imprisoning you in the fake news bubble finally is forced to admit to their deception and communicate with you directly in order to avoid having their own sanity eroded by the fact that you can bend their subjective reality in a retaliatory way, thus bringing them to the bargaining table.

How you can exploit this

Everything that we interact with in our life is made up of different *frames*, right? For example, consider the sentence "I didn't say I stole her purse." This sentence can have a wide variety of different meanings depending on which individual word is emphasized to *frame* the sentence.

I didn't say I stole her purse = I didn't say that, Becky said it

I didn't say I stole her purse = I didn't say it, I wouldn't say it, I categorically deny it

I didn't say I stole her purse = I didn't **say** it, but you know, I kind of **implied** it

I didn't say I stole her purse = I'm not the thief, it was that bitch Becky! Stay in your lane, Becky!

I didn't say I stole her purse = I have her purse, but it wasn't **stealing**, it was a gift

I didn't say I stole her purse = I steal purses all the time, but I'd never steal **Becky's** purse! She's good people

I didn't say I stole her purse = Bitch I don't steal **purses**, I steal **hearts**

The point of this is that by subtly emphasizing different aspects of a sentence, we can change its entire meaning even though the sentence itself is exactly the same. In a similar way, we can "spin the news" while still being entirely truthful simply by putting a subtle emphasis on different **parts** of the news. With the use of algorithms, this can be done in real-time as a sort of filter. The target will know about exactly the same events as everybody else around them, but those events will be **framed** entirely differently because they are filtered with a different spin, causing the target to live in a bubble of "fake news." For example, take the hilarious fact that there are Wikipedia articles about Taylor Swift, but there also exists an alternate universe "Swiftipedia" version of these articles – basically an elaborate bubble of real news that is *framed* in a way more conducive to Taylor's personal reality. Now imagine how effectively something like that might operate if you **amplify** the effect 100 times using GPT-2 algorithms, or maybe the next gen – GPT-3.

Chapter 5
What Fate has In Store
How to Avoid Bad Outcomes

THE PATH WE ARE CURRENTLY ON

Now that we know that the future of the world is both predictable (using game theory) and changeable (using memetics), it seems obvious that our next logical step is to predict it and change it. This chapter is all about that, which is why it may be a bit depressing to overly idealistic optimists. At the present moment, the course that we're currently set upon looks very bleak. A lot of bad news is going to go down in this upcoming decade, and in its current state, I don't think human civilization is going to survive this. The good news is that to Dark Rationalists such as yourselves, the future is not set in stone. It is totally changeable, which means that in theory not only can you save the world, but also have a lot of fun and advance your own personal goals while doing so.

However, before we can steer the future, we must first examine it honestly to see what course we are currently on; looking at our trajectory not through the lens of wishful thinking and self-delusion, but through evolutionary psychology and game theory. In other words, assuming that we do nothing to change the topography of the incentive structure that we currently live in, how does this all end for us? Let's talk about the unfortunate consequences of thoughtless behavior.

Obviously, prophecy is not an exact science (yet), so before we continue, I want to offer a few disclaimers. First of all, I'm doing all of these calculations in my head. We don't yet have the software tools needed to map out incentive topography in a rigorous way, and given that I'm the first modern-era person to successfully map out human behavior to mathematical logic, my models are definitely far from perfect. They could use some tweaking, and in the coming years I think that a lot of money and effort will be put into refining these formulas. For all intents and purposes, this is a brand-new science, so I think it will be a high-growth field. However right now the field is still in its infancy, so there will always be a certain margin of error. I will explain the principles behind each prophecy and my reasoning for why these principles will interact to make the future unfold in a particular way, but if I'm off on something by a few years, don't bust my balls about such trivial deviations from trajectory. Instead consider that visualizing incentive topography in one's head is *hard work* and most people would say that it's pretty impressive to be able to predict the future in any detail at all.

Second, now that you have the tools with which to reshape the future, I fully expect you to do so, because you're not stupid and I think we can all agree that a future that destroys all of us together is not particularly desirable. These mental tools are *useful* and it is my expectation that you will most definitely use them. So as I "prophesy" how Moloch ends the world according to Game Theory, I hope you will take careful notes about how to use memetics to disrupt his "plans." The whole point of discussing the end of civilization is to give you a way to stop it from ending, since it would kind of be what the kids call "a total dick move" to tell you about something like that if you couldn't stop it. Because of the butterfly effect, the further out in time we look, the more opportunity you have to disrupt the flow of events, and I plan to give you some ideas to do so. Such disruption ought to be clearly noticeable; now that you know what these techniques look like, it should be fairly obvious when somebody else is using them even if you cannot pinpoint the exact source.

THE DEFECTORS DILEMNA

Instead of beginning with what we are doing right when it comes to addressing civilizational collapse, it may be easiest to first discuss what our neoliberal world leadership is doing **wrong**. Our philosophy when it comes to issues such as global warming is deeply dysfunctional because we are not only asking the wrong questions, but we are also operating on some fundamentally misguided assumptions about human nature. The combination of these incorrect assumptions leads to defective policy.

One question that our leaders often ask themselves is how we can stem the flow of pollution. This is important because as our planet's population grows, the environmental costs to our planet grow as well, especially as countries modernize and gain an increasingly large carbon footprint. Indeed, if every country in the world were to industrialize to the level of the United States, our ecosystem would collapse in just a few decades, leading to mass death, civilizational collapse, and the gradual extinction of humanity.

Personally, I feel that the question itself is framed incorrectly because it treats population growth as both inevitable and justified. The earth can absorb a certain amount of carbon each year, so as long as we keep pollution below the threshold where the planet can recover from it, then we don't have a problem. By contrast, population growth will **always** be a problem. Even if we hit zero emissions through some miracle of technology (this will not happen, but let us assume it could simply to steelman our leadership's position) population growth will eventually lead to resource constraints that will cause famines, plagues, wars, and other unpleasant side-effects. So here is my question to our leadership: why are you focused so much on pollution rather than the underlying root cause, which is overpopulation? Why should the rest of us change our lifestyles so that some irresponsible person in a third-world country can pump out six children, even though that is totally unsustainable for the planet? Maybe we should be more focused on changing the irresponsible person's behavior instead of our own. I don't think having six children is a fundamental human right. I think that our planet's resources are limited and if some people insist on using more than their fair share – whether through pollution or overpopulation - we are justified in killing them and taking back the shared commons that they are exploiting through their selfish behavior. After all, their short-sighted and selfish behavior is killing us: why should we hesitate to retaliate in kind? If you're on a lifeboat in the middle of the ocean with somebody and they're eating all the rations and trying to start a campfire in the middle of your raft, that's dangerously crazy behavior and no jury would convict you for knocking them out and sending them to Davy Jones's locker… so why should we be expected to behave any differently when the lifeboat is our planet and the dangerously crazy behavior on display is unchecked population growth?

Naturally, killing is always a last resort. There are a lot of far easier and more subtle ways to prevent overpopulation, none of which have been implemented by our incompetent global leadership. For example, it has been conclusively proven that educating women and providing them with economic opportunities slows the rate of population growth, because women choose to delay having children until they are able to provide for them. That suggests that the best way to limit overpopulation is through financial aid packages towards developing countries. In fact, we currently offer a substantial amount of these financial aid packages, but historically they seem to have had very little effect. Many taxpayers who are indirectly funding these aid packages may wonder exactly what the problem is.

The problem that our leadership doesn't seem to understand (or doesn't **want** to understand) is that these financial aid packages are not being used appropriately. Africa and the Middle East are what we call low-trust societies: places where endemic corruption and nepotism mean that a large proportion of economic stimulus packages are misappropriated by corrupt bureaucrats and government officials. In other words, it doesn't matter how much money we spend on aid packages to these developing countries if most of that money doesn't reach its intended destination. The governments of these areas are not incentivized to solve their problem, because for the most part, they don't **care** about the well-being of their people. All they care about is lining their own pockets. In fact, I would go so far as to say that many government officials in these third-world countries actively want to **avoid** solving the problems of poverty and illiteracy among their people, because if those problems were solved, the financial aid packages might stop, and then how would these officials skim off the top? What I'm trying to say is that until their corruption problems get fixed, sending financial aid to these countries is like giving money to a crack addict. We are not solving their problems by giving them money; we are just acting as enablers and making their problem worse.

Of course, Game Theory suggests that there is an easy solution to the problem of corrupt government officials: it is called fear of death. The high-trust societies of this planet currently have a military engine quite adequate to rolling over the low-trust societies as easily as a steamroller over pizza dough. We could replace any of these governments and enforce better behavior simply by monitoring and punishing government officials that pilfer from financial aid packages designed for the benefit of their people. Some U.N. officials might say that this is unrealistic; that even though we can easily conquer these societies, we would be unable to maintain control over them due to insurgencies using guerilla warfare tactics against us. I would counter that our current advances in AI have made it easy to design robotic warfare units that specialize in eliminating insurgencies. It's fairly obvious to anybody in the know that we have the technology already to make some terrifyingly lethal automated hunter-killer units absolutely perfect for counterinsurgency operations. The only thing stopping us from applying these engineering innovations to solve the problems facing our modern world are U.N. policies that prohibit research in automated warfare. In other words, not only does the U.N. enable corruption and poverty in low-trust societies, funneling money to corrupt bureaucrats instead of the citizenry, but they also prevent research into any technology solutions that could resolve the problems that they are unable to. Doesn't that seem like deeply incompetent leadership? Fortunately the U.N. is functionally paralyzed due to the structure of their organization, which means that they are already halfway down the path to complete self-destruction. All we need to do is give them a little nudge to go right over the tipping point. With the U.N. gone, the more competent world powers can create a new global organization dedicated to getting results, without the delusion that low-trust countries full of nepotism and corruption – countries whose government officials have utterly failed to serve their own people - deserve a say in their own governance. We might even dispense with the illusion that these countries have self-governance at present – from where I am sitting many of them look more like tinpot dictatorships that are democracies in name only, and it seems to me that we would be doing the citizens of these countries a real favor by wiping out the corrupt rulers that currently oppress them. We can literally create a worldwide utopia, and all it will take is a little elbow grease and acceptance of modern technology solutions.

The other obstacle to solving the world's problems is guilt over colonialism. You see, whenever anybody mentions that perhaps our black brothers and sisters are not being well served by living in societies where rape and cannibalism are daily facts of life, and suggests that perhaps the civilized countries of the world should take military action to replace these leadership of these underdeveloped nations so that Africa can culturally improve itself, they inevitably get accused of racism and colonialism. Personally, I think it is far more racist to allow our fellow human beings to live in such sick societies when we have the power to easily improve their living situation, simply because our world leaders want to avoid putting in the *hard work* that would be required to fix these problems. Apparently wanting to improve the world by holding inadequate leaders accountable for their failures is now racist, because progressive dogma and postmodernist nonsense have degenerated our society to the point where words have lost all meaning. Perhaps this is the reason why progressives use censorship and public shaming to silence their critics: because if they didn't, people might be tempted to point out the proven fact that while colonialism admittedly was terrible and brutal in the short-term, virtually every single society that suffered under colonialism had better long-term outcomes than those which did not. I have often heard progressives claim that they care about science, which is why it is so curious that they work so hard to suppress scientific facts and data which go against their narrative.

The fact that these urgent global problems have not been solved – even though the solutions are so blindingly obvious – is an example of something called the Defector's Dilemma. The Defector's Dilemma essentially states that on aggregate, people will generally defect from shared rules of cooperation to do what benefits themselves. For example, intellectual property laws encourage innovation, thus advancing technology. China ignores IP laws because it is cheaper to steal than to innovate. End result: the world becomes a slightly shittier place because technology advances slower, but China is better off. Antifa ignores laws about violence because it is easier to achieve political goals through fear and intimidation than by working to sway hearts and minds. End result – society becomes a slightly shittier place because politics is more violent, but Antifa is better off. If aliens ever came down to visit our little blue sphere, I suspect that a crack team of ~~thieves~~ scientists would be assembled to learn as much about their technology as possible, even though it would be far more wise from a diplomatic standpoint to **pay** the aliens for their technology instead of **stealing** it. End result: the ~~thieves~~ scientists gain a short-term technology advancement, at the cost of completely ruining any chance of a positive relationship with a more advanced species.

Fortunately, the Defector's Dilemma is fairly easy to solve. Game Theory states that the optimal way to eliminate defector behavior is to punish it vigorously, a strategy known as tit for tat. For example, imposing significant economic costs upon China until their economy collapses (leading to violent uprisings against the ruling class) is a game-theory effective method of forcing China to play by the same trade rules that everybody else uses. Encouraging far-right groups to commit anonymous violence against Antifa supporters is a highly effective strategy against Antifa violence because once Antifa realizes that violence against others results in massive costs to them, it becomes more cost-effective to them from a game theory perspective to start playing by the shared rules of civilized society. And if our hypothetical aliens ever asked me for advice on how to deal with technology theft, I would advise them to read up on Roko's Basilisk and start sending the ~~thieves~~ scientists ~~souls~~ data to a simulated equivalent of hell for eternal torture after they die, while creating a simulated version of heaven for people who cooperate with them.

In all of these situations, the Defectors Dilemma is easy to solve simply by pre-committing to punishing defectors so harshly that it is far better for them to cooperate rather than defect. The reason great empires collapse is because they forget this lesson. Harsh methods are needed to carve civilization out of anarchy (and keep it from falling back into anarchy), but with enough time spent living in the light of civilization, subsequent generations grow soft and forget the necessary of punishment, subscribing to the deluded idea that human nature is fundamentally **good**. When the public buys into a Great Lie of that magnitude, it changes everything about how we interact with the world, shifting society into a state where collapse is inevitable. The only way to fix society at that point is to expose the lie, thus shifting society back into a state that is more game-theory optimal. The truth is that human nature is neither good nor evil – it simply follows the same incentives as that of all Life. If you make it more optimal from a game theory perspective for people to do the right thing and cooperate with each other, they will do so, and if you make it more optimal for people to prioritize their own selfish short-term interests over the well-being of the shared commons, they will do that instead. Personally, I prefer a world where people do the right thing and cooperate with each other, but that means that we need to collectively acknowledge the need to punish defectors, in order to make it more game-theory optimal for people to cooperate rather than defect.

OVERPOPULATION AND SOCIETAL COLLAPSE

Let me ask you a hypothetical question. What would happen to a society if half of their women decided to just **leave**?

For starters, their birth rate would drop significantly. That's a no-brainer. Less women equals less children. Another thing that would certainly happen is that their crime rate would go up. Our politically biased sociologists may be wrong about a great many things, but one thing that they are most definitely correct about is the fact that most violent crime is caused by men. In fact, I would go so far as to say that women are a significant stabilizing force in society, and data backs me up on this. Once the ratio of men to women exceeds a certain point, social unrest becomes a significant problem. So why have no governments weaponized this yet? (Some might say that many European governments have actually weaponized this against themselves, by importing refugees who are predominantly young men, but generally speaking "weaponizing" a tactic refers to using it against geopolitical rivals rather than yourself.) All the societies that seems to be causing the most headaches for the rest of the world – both in terms of human rights issues and overpopulation problems – are culturally backwards patriarchies that treat women like second-class citizens. If we allowed women from those societies to immigrate here freely but prohibited their men from immigrating, not only could we destabilize hostile societies, but also help solve the overpopulation problem in a single stroke. The fact that these women are often legitimately in fear for their lives would also offer a convenient smokescreen for these policies, allowing us to gain the moral high ground by turning this into a human rights issue. Some people might say that it is unethical to deliberately collapse enemy societies by rescuing their women from a patriarchal culture of abuse and oppression, but I think nothing could be **more** ethical. Have you seen the data on the human rights abuses that go on in some of these places? The way women are treated in many of these cultures is truly shocking. It's interesting that American feminists seem so concerned with "microaggressions" when their counterparts in other countries have "macroaggressions" to deal with; almost as if third-wave feminism is really more about narcissism and political grandstanding rather than actually helping women. But perhaps that's a bit cynical of me to suggest. We can revisit that topic later.

What's another way that we could deal with overpopulation? Artificial intelligence, for starters. I've already discussed some of the various military uses of AI, but did you know that one of the biggest "early adopters" and drivers of technology is the porn industry? That's right; I'm talking about sex robots. In laymen's terms, people like to have sex. One type of sex **often** results in people getting pregnant, and the other type **never** results in people getting pregnant. By making the second kind of sex much more attractive than the first kind, we can easily solve our overpopulation problem. Some might argue that even though we can easily make robots physically more attractive than people, AI can never really mimic the true depth of a human personality. To which I would answer, have you stepped outside lately? We live in the most vapid, shallow, narcissistic culture that has existed in centuries. We could program a robot to only have three or four default responses and it would **still** be far more interesting than many of the people I've met on dating websites. I really don't think AI is going to have much trouble replacing us in the sexual marketplace. The only question is whether it will be allowed to. It seems that once again, third wave feminism is getting in the way of developing technologies that could save our planet, making the same tired old "slippery slope" arguments that so many self-righteous fools have used in the past.

It's interesting that identity politics keeps on throwing up arbitrary obstacles to solutions that could save our species from certain death. Could it be that instead of being an altruistic philosophy designed to foster equality and tolerance, identity politics is really just a naked power grab, designed to gain a short-term advantage at the expense of long-term solutions? That sounds exactly like the definition of the *Defector Dilemma*, a basic fundamental principle of group behavior! And the way these people support whatever platform personally benefits them most and then come up with creative rationalizations to justify their self-centered behavior is just like those hypnosis experiments I mentioned at the beginning of chapter 2. *Cognitive dissonance*, another fundamental principle of group behavior.

It's no coincidence that these basic principles keep recurring over and over in everything we do. You see, the reason that economists, psychiatrists, and philosophers have always had such difficulty predicting group dynamics is not because their models are too simple to handle the complex nuances of human behavior: it's because their models are too complex and human behavior is really quite simple. The only complex parts of our thought processes are the delusional mental gymnastics we use to rationalize our behavior so that we can create a mental narrative that frames us as being good people. Eliminate that self-delusion, give people a peek behind the Great Lie, and suddenly forecasting becomes very simple. As I told you at the end of Chapter 1, we all have the innate ability to see the future. We subconsciously **choose** not to do so because it would force us to accept the unpleasant truth about whom we really are. 99% of your personality – what psychiatrists call the ego – is self-delusional bullshit meant to create a heroic narrative in your own mind, allowing you to see yourself as the protagonist of your own story. Strip that away, and what is really left? What is left of you, once we subtract all the lies that you tell yourselves? The sad truth is, very little. But this is a necessary step in personal growth because it is only once we have deconstructed our fragile personalities that we can begin the process of reconstructing ourselves in a more rational and resilient way.

But we're not there yet, and if identity politics has its way, we never will be. So let's talk about the course of history if we keep going along the same "compassionate neoliberalism" path that we have been operating under for the past several decades.

First of all, the problems in the third world will never be solved, because our leaders will continue to fund third world dictators who oppress their own people. Our leaders won't **admit** that's what they're doing, of course – they'll show you tragic pictures of starving people overseas and sad little children in detention camps (and call you a monster if you refuse to let your rational thinking be overwhelmed by the way they cynically try to manipulate your emotions) but that's effectively what's going on. When you examine the amount of foreign aid we give to third world countries versus the amount that is actually used for its intended purpose of helping their citizenry, it's just a trickle. Where does the rest go? Why, to consolidating the dictatorship's power, of course! They need guns and ammo in order to properly oppress their own citizens, not to mention building rape dungeons for their depraved offspring. The next time you see some bleeding-heart politician making a passionate speech asking for more funding and development aid to the third world, remember that what he is really talking about is building more rape dungeons. He won't admit it to you – in fact, he probably won't admit it even to **himself**, because I imagine it must be very difficult to look yourself in the mirror each morning knowing that you work for a "peace" organization so incompetent that it actively helps support such atrocities – but at the end of the day, that is what he is talking about, because when you send development aid to third world dictatorships without some way of violently enforcing the way that the development aid gets spent, it inevitably gets spent on oppressing the citizens of that country. Just because you **meant** well doesn't matter; the outcome is still the same. You might not like to think about it that way because it would mean that many people who are supposedly the world's most compassionate leaders actually have a lot of blood on their hands, but you bought this book because you wanted to peek behind the Great Lie, remember? Please don't blame me just because you don't like what you see.

But as long as we're talking about blame, let's not assign too much to any specific politician, because there's plenty to go around. You're equally as much to blame, because anytime somebody tries to put in the *hard work* of replacing these dictatorships and creating a more stable and less corrupt government in their place, a lot of left-wing zealots obsessed with identity politics start accusing them of colonialism and nation-building. Yes, my distinguished colleagues, that's exactly the **point**. These people don't have very good nations, so we need to colonize them and rebuild them to make them **better**. The citizens of these countries can't do it on their own, because thanks to all your well-intentioned "development aid," their corrupt governments have all the weaponry and their citizens have none. You and all your misguided good intentions turn these countries into barbaric hellholes, and then when some politician with a brain and analytical skills actually tries to fix the root cause of this problem by going in hot to take out the dictators at the top of the power structure who are causing all the problems, a bunch of clueless hippies start chanting about "giving peace a chance," totally undermine the war effort, and cause the whole push to civilize these places to collapse. So while we should certainly blame all these politicians for their hapless incompetence, perhaps it would also be wise to do some self-reflection and question whether we are giving them the right *incentives*. Systemic breakdowns are seldom caused by just one person. They are caused by a domino effect: a long chain of events that starts when people stop doing the *hard work* of punishing defectors – whether those defectors are at the top of the food chain or the bottom.

Although it is obviously sad that corrupt governments oppress their own citizenry (with your own politicians' tacit approval, of course), that level of abuse is only a local problem, and I want to talk about how this ties into our global problem. Do you remember earlier in this chapter when I mentioned that educating women and providing them with economic opportunities was proven to slow the rate of population growth? Well, what do you think happens when all that "development aid" money that you **thought** was going towards women's empowerment instead ends up being misappropriated to fund some dictator's private army, because your politicians decided to give them money without any means of enforcing how they spent that money? That's right, the population of that country grows rapidly. And obviously you can see that when you have a third world dictatorship suffering from an overpopulation crisis, there aren't enough resources to go around. People end up living in extreme poverty. This relationship is so consistent that it's almost like an equation.

$$quality\ of\ life = \frac{resources}{population}$$

As terrible and tragic as this is, one of the bright sides about extremely poor societies is that if you leave them alone then over the course of time they tend to be self-regulating, in the sense that extreme poverty is not a static equilibrium point. If you have enough poor people living in a country under a corrupt and exploitative government, eventually the pressure builds to a point where it explodes into revolution and the poor people end up overthrowing their government. All you have to do is leave the situation alone and avoid taking pressure off the dictator by accepting refugees, because each additional poor person in that corrupt regime increases the pressure on everybody else, and once the average quality of life dips low enough, a violent revolution against the government is inevitable.

$$quality\ of\ life * X = \text{time to next revolution}$$

So obviously if you want to take out a corrupt and abusive dictatorship without lifting a finger, one very important thing that you need to do is impose immigration restrictions upon them. Don't let people out of that country, because the more people you cram into that dictator's country, the more of a pressure cooker it becomes, until it explodes violently. But of course, our ignorant governments do exactly the opposite of this: they encourage immigration from authoritarian regimes in the name of "compassion," thus perpetuating the existence of these third world dictatorships by offering them a safety valve. To me, it doesn't seem very compassionate to consistently take actions that would leave the world in a state of backwards savagery – on the contrary, it looks like the kind of thing that a coldly manipulative sociopath would do to virtue-signal that they were a compassionate person, because over the long term it causes far more pain and suffering than it would if we simply sealed the borders and allowed the pressure cooker to explode. But humans didn't evolve to be kind or cooperative, they evolved to **persuade** other humans that they are kind and cooperative, and the best way to sucker other people into believing this lie is to make it so persuasive that you also convince yourself. If you ever wondered why our societies seem so dysfunctional at times, or why our planet is heading for total environmental annihilation even though all of our leaders claim to care about this issue intensely, there is your answer: it's because most of our societies are run by a bunch of narcissistic sociopaths who waste 99% of their brainpower rationalizing their actions to convince themselves that they are actually good people. In other words, they're not just evil, they're **also** stupid. Evil people at least can be effective leaders if you can arrange their priorities in such a way that they align with the interests of society, but evil **and** stupid is a toxic combination.

At this point you may have some reasonable questions, like "What was that variable X that you mentioned there? Unknown variables make me nervous!" As well they should! When you're trying to calculate an equation and the result impacts entire societies, unknown variables should make anyone nervous. The short answer is that X is the special sauce. We don't fully know yet what ingredients go into X, although we can make a few educated guesses. For example, the fact that a higher man to woman ratio destabilizes a society while having a higher woman to man ratio makes it more stable is definitely a component that would factor into whatever equation governs X.

Some might say that because understanding the equations that govern group behavior is so vital to building a better world, it is very important to research and develop this field more. Unfortunately, this is currently very problematic, thanks (once **again**) to identity politics. You see, because game theory is based on evolutionary psychology – a field which is not very "politically correct" according to today's orthodoxy - it is very hard to study the field without coming to the conclusion that imperialism is both necessary and justified. Furthermore, it is empirically provable through the mathematics of incentive topography that if white people had not enslaved blacks, then black people would have enslaved whites… which obviously puts some uncomfortable pressure on the narrative that our media currently seeks to push forward. Obviously, these are all very controversial statements because they completely invalidate the grievance narrative that a lot of high-status people in our society currently exploit for political and financial gain. Even though everything I've said here is empirically provable, our perpetually outraged media outlets and politicians would falsely brand me a racist for making this claim, and would try to destroy my reputation and career, just as they have done to so many other innocent people. Another way to frame this is that because people in today's societies lack the ability to mathematically calculate the outcomes of their actions, they have a very deluded sense of right and wrong. Giving them the ability to calculate incentive topography would require many people to admit that their existing ethical systems and beliefs result in bad outcomes, and a lot of particularly angry and self-righteous people would have to admit that they have actually been wrong the whole time about practically everything. A lot of power is also likely to change hands as a result of this, because once it can be mathematically proven that identity politics has bad outcomes, then many politicians, entertainers, and executives whose careers have been rooted in identity politics are going to very rapidly find themselves called to account for their bad decisions. In other words, many of the people who have a vested interest in the existing system are going to fight tooth and claw to preserve the status quo, even though it is bad for everybody else and will ultimately lead to the downfall of civilization. This is not new behavior: it is a truism that science has always advanced one funeral at a time. As Upton Sinclair said "It is difficult to get a man to understand something when his salary depends upon his not understanding it." So perhaps we shouldn't even bother trying to get these selfish people to understand. Perhaps we should be more focused on getting everybody **else** to understand, and then once we have a clear majority on our side, we can all work together collaboratively to take these selfish people out. I for one have no intention of letting our planet die slowly just because some self-centered elites don't have the humility to admit that they were wrong and made bad decisions which hurt all the rest of us. How do you feel about it?

CONFLICT THEORY VS MISTAKE THEORY

As observant readers, you have undoubtedly noticed that Scott Alexander is one of my favorite rationalist bloggers. But even the smartest people make bad mistakes sometimes, and in this section I need to talk about the most significantly flawed part of his philosophy, which is summarized in a post entitled Conflict vs. Mistake. In this blog post, Scott summarizes the difference between mistake theory and conflict theory. The post is well worth reading, but to summarize it briefly, mistake theorists assume that when people have a disagreement about what policies would best serve the public trust, it is largely because of a genuine disagreement about what would be best for the public. In such a worldview, the best way to settle the disagreement is through reasoned discussion and debate. Conflict theorists assume that when people have a disagreement about public policy, it is because different groups have different vested interests and they tend to assume whichever position would benefit them most. In the worldview of a conflict theorist, the best way to settle the disagreement is through conflict. Scott concludes that he tends to fall on the side of mistake theory, which is frankly why he has been so remarkably unsuccessful at spreading rationalist teachings – because mistake theory is entirely wrong. The truth is that people operate along conflict theory, but then spend a remarkable amount of time rationalizing their behavior so that they **think** they operate along mistake theory. This process, which we refer to as *motivated reasoning*, allows people to be totally selfish while at the same time creating a heroic narrative to help them convince themselves (and more importantly, convince others) that they are good people. "Good people" get more reproductive opportunities and tend to advance faster in group hierarchies, so at the present moment in time this strategy of self-delusion is highly *useful*, which is why we evolved it. Everything that we believe – every ethical system or social custom that we have – only exists because said behavior is somehow beneficial to achieving Life's goals. Once such behavior stops being adaptive, or a new ethical system appears that outperforms it, outdated ethical systems tend to die very quickly, along with whatever Life still clings to them. This is how belief systems evolve over time (a topic that we will revisit very shortly).

   Let me give you an example of how *motivated reasoning* works. One of my former heroes, a Silicon Valley visionary named Peter Thiel, is a passionate advocate of cryptocurrency, the most famous of which is Bitcoin. Understand that from the perspective of "what is good for society" there is absolutely no **good** reason that cryptocurrency should exist. From an environmental perspective, it is terrible, and the process of "mining" cryptocurrency uses more electricity (generating a corresponding amount of pollution) than some small countries. From the perspective of human rights, cryptocurrency is primarily used to finance illegal transactions, such as the drug trade and human trafficking. From a governmental perspective, crypto is the currency of choice for evil despotic regimes, helping to prop up governments too corrupt and incompetent to maintain a healthy economy. Peter Thiel is a very smart man, so it's not as if these problems haven't occurred to him. The **real** reason he thinks cryptocurrency is a good thing is because it will make him and his Silicon Valley buddies very, very rich.

   Of course, Peter Thiel can't **admit** this to himself. If he did, he would have to acknowledge that at his core, he is in some ways a selfish person who is willing to contribute to human trafficking, despotic regimes, and the destruction of our environment, all so that he can make money. This is where *motivated reasoning* kicks in. Thiel's subconscious mind knows that making cryptocurrency succeed would be advantageous for his goals, so it wastes a lot of valuable processing power coming up with a persuasive story about how bitcoin is somehow good for humanity because it "liberates us from the tyranny of fiat currency" or some other similar-sounding nonsense. Once Thiel's subconscious has convinced him that making cryptocurrency succeed is not selfish – that it is a good, even **noble** decision – then Thiel can fool himself into thinking that the selfish advancement of his goals is somehow a heroic quest. More importantly, he can fool others into believing the same thing, since it is far easier to convince somebody of something when you genuinely hold that belief yourself (which is the reason that our minds evolved the capacity for self-delusion in the first place).

Is Peter Thiel a bad person? Based on how successful he has been, evolution would tell us that he is not. A lot of very wealthy people like to bullshit themselves about how virtuous they are, so from an evolutionary perspective self-deception is a very beneficial cultural adaptation. It only dies out when a more successful memetic adaptation comes along, rendering it obsolete. For example, when an ambitious author points out that by stripping ourselves of this delusion and practicing radical honesty and self-awareness, we can literally learn to predict and shape the future, a far more *useful* talent that makes self-deception obsolete. Personally, I like Peter Thiel, but I think that like many people he could stand to benefit from a little more awareness of his own subconscious thought processes.

What *motivated thinking* means – in purely practical terms - is that the best way to convince people of something is not through logic; it's through firepower. If somebody is making a lot of money from polluting the environment, Scott Alexander could show up at their door with a million good reasons and solid arguments why polluting is a bad thing, and they would scoff and find a way to dismiss him, most likely by calling him a bigot. But if I showed up at their door with a million good people and solidly explained that we were collectively going to pound them and everyone they love into a greasy smear unless they stopped polluting by the end of the year, not only would they do it, but they would eventually convince **themselves** that polluting is bad, and that the reason they stopped was not because of my threats, but because they had "seen the light" and genuinely wanted to do the right thing. With enough time (and a little assistance) they might even rewrite their own mental narrative so that they had **wanted** to stop polluting all along, and me and my million followers had simply given them the gentle push they needed to admit it to themselves.

My point is that if you want to genuinely change hearts and minds, it is pointless to try to engage or debate with the *heroic narrative* that Life creates. Most of the opinions that we hold and the narratives that we create for ourselves are entirely the product of *motivated reasoning* and are totally fictional. The best way to change hearts and minds is to completely ignore people's stated rationalizations for behaving the way that they do and instead engage with the subconscious rationalization processes that actually decide what Life wants – the *incentive topography* of said Life. In other words, if you make it clear that helping you will result in lots of reward and opposing you will result in lots of punishment, Life will do what you want and then convince itself that this is what it actually wanted to do all along. Another way to think about this is through the lens of incentive topography. Most people, no matter how smart they are, take the easy path in life - or at least avoid the hardest options. For example, Scott knows that a lot of controversial rationalist beliefs are true, and a lot of politically correct beliefs are totally false, but because he lives in the Bay Area monoculture where saying such things would make his friends ostracize him, he avoids publicizing these harsh truths and tries to soften what he says in a way that is less "offensive." Moreover, he convinces himself that he is a good person for doing this, and that such intellectual dishonesty is somehow justified. This lesson here is that how smart somebody may be is completely irrelevant to their capacity for self-deception. Very few people are capable of looking at the world without blinders on, because under the current paradigm, self-deception is a *useful* trait. The purpose of this book is to teach you that self-knowledge can be even more *useful*. Now let's talk about how to use that knowledge in order to change the world.

HOW TO WEAPONIZE KNOWLEDGE

The first step in changing the world is to understand who runs it and why. Specifically, the world is run by groups of wealthy elites who generally acquired their wealth by exploiting the existing paradigm – a paradigm where self-delusion and virtue-signaling are rewarded. Because of this, they tend to be very resistant to any attempts to change said paradigm, and will mobilize substantial resources against anybody whom they perceive as a threat to that paradigm. This means that the system tends to have a certain inertia. To overcome this inertia, you need two things: allies, and resources.

Knowledge is very helpful in acquiring both of these things, because Life tends to be grateful when you do something that helps it achieve its goals, and knowledge is very *useful* in that regard. In laymen's terms, people evolved to be cooperative, so if you offer somebody an easier way to get what they want out of life, they will tend to like you and will try to help you get what you want out of life. Offering them a way to get resources and power more efficiently is doubly advantageous, since most Life likes having resources and if you give people useful advice for how to harvest resources more effectively (for example, by having an entire chapter in your book dedicated to teaching people how to subvert political structures and financial systems in order to reposition themselves at the top) then people who take your advice and make money from it will tend to like you. You have thus solved the problem of gathering both allies and resources in a single stroke. This means that when elites with resources inevitably try to steamroll you, using the usual approaches (false accusations of treason, racism, pedophilia, murder, theft – whatever they think they can try to stick you with in order to discredit you in the eyes of the public) you will have a lot of solid well-financed allies to back you up and hit back against your enemies. And if you demonstrate that you're willing to play the game just as dirty as they are, and have a much larger toolbox to help achieve your goals, then resistance to your objectives has a tendency to just melt away, as people who work with you become rich and successful, while people who work against you gradually die out due to the pressures of Social Darwinism. When it becomes clear that your allies can win favor with you by attacking your enemies – and that in fact you actively encourage this – then as your allies gain more power and resources, people will start to show an increasingly large hesitation about antagonizing you.

But this is all theory, and the purpose of Dark Rationalism is to find practical real-world applications of theory (which is the reason I like to give examples whenever possible). So here is an example of how this might work.

I believe that the fields of economics and sociology are both garbage sciences which are 90% politicized bullshit. I have tested some of their hypotheses and they do not check out at all. Furthermore, the complete inability of economists and sociologists to reliably make accurate testable predictions is one of the classic hallmarks of bogus pseudo-science. I believe that Game Theory and Memetics are far more accurate predictors of group behavior, and I think that we need to completely discard economics and sociology in favor of these new, more accurate sciences. Much like astronomy, phrenology, alchemy, and all the other pseudo-sciences that preceded them, it is time for economics and sociology to either evolve or die.

If I were operating along mistake theory (as Scott Alexander wants us to do) I would politely advance my argument in a scientific journal, at which point all the leading economists and sociologists would laugh and ridicule me, because acknowledging that I am right would result in all of them losing their jobs. If I persisted and it looked like my ideas were gaining more serious traction, then my opponents would escalate to smear campaigns and harassment. The real reason they would do this is because my ideas pose a threat to their legitimacy, and suppressing the truth is the only way that they can maintain their status and power. However, due to the phenomenon of *motivated reasoning*, their subconscious minds would come up with some way to rationalize their behavior in order to maintain the fiction that they are still the good guys, despite their dirty tactics. I imagine that they would convince themselves that I posed a threat to "global stability" (ie, the entrenched system that they benefit from) and so any underhanded tactics they used would thus be justified, because the end justifies the means and all that.

Fortunately, I am not operating along mistake theory but along conflict theory, which paints a far more accurate picture of human behavior. In other words, I knew from the start that it would be impossible to convince sociologists and economists of this truth, so I decided to simply bypass them and convince everybody else. The reason that an entire chapter of this book is dedicated to practical methods of allowing you to seize money and power and claw your way to the top of various social hierarchies is because I want you to actively go out and **prove** the value of my techniques. The status of scientific and academic elites comes from their credibility. Once it becomes clear that my techniques are superior to theirs, their credibility is gone. At that point, they probably get fired, since nobody wants to pay lots of money for garbage pseudoscience when far more reliable techniques are available. This means that the money and influence which were previously routed toward economics and sociology "experts" get rerouted to memetics and game theory experts instead. While mistake theorists like Scott are bleating in the wilderness, begging for their intellectual inferiors to pay attention to them and give them the respect that they are due, conflict theorists like myself prefer to demand respect, taking it through any means necessary. In my opinion, mistake theory is all about "starting a conversation" which is counterproductive because "conversations" somehow never seem to resolve anywhere except the existing consensus. In other words, if you're right and the dominant scientific consensus is wrong, under mistake theory you get laughed at and people ignore you. In terms of scientific and cultural advancement, conflict theory is far superior to mistake theory because when you weaponize science and culture, then it doesn't matter what the existing consensus is, it only matters who is **correct**. If you are correct and the existing consensus is not, then the existing elites can either switch to the correct paradigm or end up humiliated and destroyed when their methodologies are outperformed by a superior *historical narrative*. This results in science and culture advancing much faster than usual because in a world where being wrong results in the loss of money, status, and power, we could say that dominant elites now have *skin in the game*.

But enough talk about society; let's talk about how this impacts you. Because more efficient and evolutionarily adaptive *historical narratives* always end up outcompeting outdated narratives, it's important to make sure you pick the right team when two narratives compete. This way, you end up on the right side of history, rather than the side that ends up getting laughed at and called backwards savages in all of the history books. Picking the right team is harder than one might think. Just because somebody **claims** to have been inspired by a new *historical narrative* does not mean that their claim is legitimate. There are an endless number of cranks and nutcases out there, all with their own unique story about how they are the "chosen one" who will guide humanity into a brighter future. So how do you distinguish between somebody who genuinely has a more accurate narrative and a charismatic wannabe cult leader? Every prophet is unique, as is every cult leader, so there are no easy answers to this. However, there are some easy questions that may help filter out noise from data, and it's wise to ask yourself all of these questions before committing to any sort of fringe philosophy or belief.

1) Can this person do things that are not explained by the current *historical narrative*? For example, things like accurately predicting the future in highly explicit ways, successfully initiating and conducting a lawsuit without any contact with either their lawyer or the defendant, crushing rock in their bare hands, walking on water, etc. If the person can manifest unusual abilities that cannot be easily explained under the current scientific paradigm, then it implies a certain degree of truthfulness to their claims.
2) Which narrative is gaining power, and which narrative is losing power? Obviously, every new *historical narrative* starts out weak, because people tend to be conformist and so they tend to dismiss new belief systems out of hand. However, since more accurate *historical narratives* are an evolutionary adaptation, you should see a general trend of the new belief system gaining steam and spreading rapidly over time. Meanwhile, the old belief system will find itself under increasing skepticism and critique. Observing trendlines like this is important when it comes to verifying the accuracy of any purportedly new *historical narrative*.
3) Which side seems smarter to you? Obviously, "smart" is a relative term, but if all other things were equal and the two sides were competing in nothing more than a battle of wits, whom do you think would win? Generally speaking, smart people tend to adopt new historical narratives before less intelligent people, so if you see a lot of really bright people suddenly shifting sides to a radically new worldview, you should consider that maybe they have some reason for doing so. If it seems like the only advantage one side has is that they have more money and power (due to being beneficiaries of the existing narrative) but the other side is still running circles around them simply out of sheer cleverness, then it's a pretty good bet that the more intelligent side will rapidly acquire enough money and power to defeat their ideological opponents in short order.
4) Which side seems more capable of problem-solving? To some extent, this may be redundant with the previous question. However, it is important to note that new *historical narratives* tend to arise when there is a systemic problem that previous paradigms have proven incapable of resolving. Examining which philosophy seems

better at successfully resolving recurring systemic problems in society is a point of view definitely worth considering.

Generally speaking, Dark Rationalism tends to offer a significant advantage when it comes to successfully identifying new historical narratives because it is a pragmatic philosophy focused on measurable testable hypotheses. The goal of Dark Rationalism is to derive a more accurate understanding of the world, peering behind the veil of all the lies and wishful thinking that our culture brainwashes us with, and this is a useful tool for separating truth from fables. That said, no matter how enlightened Dark Rationalists may be, they are only fallible humans, and thus prone to error. Therefore it is important to always apply a bit of skepticism to wild claims, rather than just going with your gut feeling. No matter how advanced in the Dark Arts you may be, you should always try to avoid overconfidence and have the humility to admit that you may be mistaken. Ultimately, the only thing you can trust is data, and even that can be faked in certain situations.

FIXING SCIENCE

In a recent book, Silicon Valley venture capitalist and libertarian advocate Peter Thiel poses a really important question: why is it that we have not made any real scientific progress in the last thirty years? By "scientific progress," I mean things like discovering entirely new scientific fields. Everything that we use today is simply a faster and better iteration of the same technology we had thirty years ago. It didn't always used to be this way. When the field of applied chemistry came into existence, it caused a bit of a ruckus, and some might even say that it changed the world. Why can't we achieve those kinds of technology breakthroughs today? Thiel spends most of his book examining this question from a variety of angles before coming to the conclusion that the problem is that we do not have enough creative and intelligent people who are willing to challenge the status quo and re-examine their own basic assumptions. While this is a very cogent observation and I wholeheartedly endorse it, I feel like it could be enhanced with a bit more root-cause analysis. In other words, let's examine this through the lens of Game Theory. We know that systems shape people rather than vice versa, so let's ask ourselves **why** the system is creating such people. In other words, what incentives are baked into our scientific process that are **causing** people to be such conformists, while simultaneously pushing out original thinkers?

We should start by defining what separates a science from a pseudoscience. For example, what separates astrology from physics? Both use complicated mathematical formulae, but obviously one of these is a real science while the other is delusional nonsense. Without having specialized knowledge about either of these fields, how can we distinguish one from the other? In my opinion, there are two distinct traits that any layman can identify to easily separate science from pseudoscience: **replicability** and **predictability**.

Replicability is a trait that indicates something happens the same way under the same controlled circumstances. For example, in chemistry if you take two chemicals that have a controlled chemical reaction when heat is applied, those two chemicals will always have that exact same reaction under those specific circumstances. You may have to set up the same environment, atmospheric conditions, heat, etc., but the end result will be the same. Under that definition, sociology does not qualify as a science (because most sociology experiments fail to replicate) while game theory does, because game theory experiments tend to replicate very predictably. In fact, from a certain perspective, some might even say that you are participating in one of those experiments now.

Predictability is a trait that indicates you have the ability to predict how something will happen under certain controlled circumstances, because the outcome corresponds to some mathematical formula. For example, in physics if you drop an object from a certain height, you can always calculate the exact moment when that object will hit the ground. Granted these calculations may be very **difficult** due to variables like air resistance, wind speed, weight of the object, and other factors, but ultimately if you have all the information it is possible to make an exact calculation of the outcome. Under that definition, economics does not qualify as a science, because economists are always making incorrect predictions. Memetics definitely qualifies as a science, because we can accurately steer human behavior in predictable ways using memetics.

While it is sad that sociology and economics are entirely fake, considering the vast amount of resources that our governments and institutions have poured into them, the situation is not entirely unsalvageable. All pseudo-sciences are eventually forced to evolve over time. Some - like astrology or phrenology – have too little value to be *useful*, and they end up dying, their practitioners ending up as disgraced charlatans in the footnotes of history. But others – like alchemy – end up evolving into real applied sciences, like chemistry. So we should ask ourselves, what would it look like if we were to turn economics and sociology into real sciences, rather than fake news? In other words, what would these fake fields look like if their research hypotheses were required to fulfil the real scientific criteria of replicability and predictability?

In terms of economics, a more scientific approach would be as follows.

1) An economist would research the existing data and develop a hypothesis. This hypothesis would fill a gap in the existing body of knowledge, or demonstrate the inaccuracy of a theory normally accepted as part of the scientific canon. For example, "The existing economic work tells us that there is a direct relationship between the supply-demand curve. My hypothesis is that under circumstances X, Y, and Z, this relationship does not hold true."
2) Based on their hypothesis, they would generate a set of specific measurable predictions. These predictions would be generated beforehand in order to verify the objectivity of the experiment. For example, "We have a few examples of circumstances X, Y, and Z coming up because of the latest EU regulations in the pipeline. The EU is operating under the current hypothesis of the supply-demand relationship and believes that this will stimulate their economy substantially, raising GDP by at least 10 percent in the upcoming year. I believe that because of circumstances X, Y, and Z, their economic predictions are incorrect and GDP will be raised by no more than 2 percent at the maximum." The reason the prediction needs to be made beforehand needed is because so-called economics "experts"

have a real knack for self-deception, and whenever they make one of their countless inaccurate predictions they always find excuses to justify why reality did not meet their expectations. By forcing them to make their predictions beforehand in a measurable quantifiable way whose results cannot be debated, we strip away their ability to make excuses and force them to compete on a level playing field where their pedigree and academic credentials mean nothing: the only thing that matters is their accuracy.

3) Time passes. Once the experiment is concluded, the predictions are measured against reality to determine how accurate they were. Economists who consistently demonstrate accurate predictions have a higher value and should receive enhanced status and pay, as well as greater promotion opportunities. Economists who consistently demonstrate inaccurate predictions should be subjected to ridicule and receive lower pay, as well as fewer promotion opportunities. This ensures that our economics "experts" have *skin in the game*, because their own success or failure is tied to how well their theories describe actual reality.

This is a fairly straightforward scientific process to enhance the effectiveness of economics, yet it may surprise you to learn that we currently use none of this methodology. Instead, economic theories are accepted into the mainstream by a process of making up elaborate stories which are then disseminated among other economists. These stories have no predictive power and instead are retrofitted to prior data. If a majority of economics "experts" agrees that the story sounds reasonable, then the theory gets accepted into scientific canon, without even any measurable tests or societal experiments to validate them.

There are multiple problems with the existing process. First of all, economists spend a lot of money getting their degrees, which means that they are reluctant to let go of outdated or incoherent theories that pertain to their field of study, since their degrees lose value if those theories are invalidated. Second, if the majority of experts are wrong, then the scientific canon is never changed. This has happened many times in other scientific fields, like astrology or alchemy (which turned out to be complete nonsense but were taken very seriously at the time), so it would be delusional to assume that we are the only group of people in history to have an inaccurate *historical narrative* in regards to our scientific accomplishments. We need to consider the possibility that there are entire scientific fields which are nothing more than pure garbage, and we need to have some sort of mechanism to flush these fields out. Third, the existing methodology offers no way for an uneducated layman to test the validity of the hypothesis. If the only way to prove that large parts of the economics canon is fraudulent is to drop $100k and several years of your life on getting an economics degree, then nothing will ever get changed in economics, because once you have dropped that kind of cash and time on an expensive piece of certification, you are maximally incentivized to ensure that the certification is valuable, and blowing up your own field of study does exactly the opposite. Fourth, since the current process prioritizes "expert" consensus over hard facts, the best way to ensure that your theories are accepted is by kissing ass to established members of the field – whose incorrect theories are part of the problem to begin with. Perhaps I see things differently from most people, but it seems self-evident to me that any science which requires one to kiss ass in order to be taken seriously is not a legitimate scientific field at all. Fifth, there is no practical value in telling stories that can only explain the past. Value is only generated by telling stories that enable us to predict the future. Nobody cares about your ability to come up with some random story to explain why the **last** housing bubble burst and caused them to lose their life savings: people only care if you can predict the **next** housing bubble in time to allow them to keep their life savings. Finally, I think that we should all frankly be embarrassed to name something a "science" when it has no predictive power and consists mainly of making up elaborate faerie-tales, cherry-picking data to support those stories, and kissing ass to pompous idiots in order to get them to put the stamp of approval on these stories. As a rationalist, this doesn't sound like science to me: it sounds like the complete opposite of science.

      Sociology has a similar problem to economics, but where economics fails the prediction test, sociology fails the replication test. The politicization of sociology is largely to blame for this. It may surprise you to learn that sociology skews between 96% to 99% liberal, and conservatives have reported substantial discrimination against them in the field. To put the immensity of this problem in context, imagine that 96% to 99% of the researchers in a field were straight white men, and that people who were not straight white men reported substantial discrimination against themselves. It would be easy to see that there was a discrimination problem – almost any court would find the institutions guilty. Furthermore, you would not expect to get good science out of such a homogenous group – there would be an obvious bias towards a straight white male perspective. Yet when we have such discrimination against ideas, why does nobody see a problem? Science is an area where diversity of thought is most needed, and indeed it has increasingly become standard practice among rationalists to study topics with a person of the opposite philosophical mindset (this is known as an adversarial collaboration). But instead of tackling the problems facing sociology with a diverse set of ideas, sociologists seem obsessed only with diversity of skin color, while working tirelessly to ensure that everybody has exactly one identical idea - the same academically approved dogma. Indeed, many sociologists report entering the field to try to promote social-justice ideas. This is problematic because social-justice ideas are often untrue. The reason human history is so bloody is because exactly 0% of our evolutionary history is concerned with social justice, and 100% of our evolutionary history is concerned with survival. When you try to interpret group behavior through the lens of whatever cultural morality is currently fashionable while avoiding any discussion of how evolution shaped our group behavior, you are ignoring thousands of centuries of data in favor of focusing only on the last one. Furthermore, when you create an experiment with a predetermined narrative that you want to push, it is incredibly easy to massage the data in a way that supports your predetermined conclusion. This is because – just like economists – sociologists do not make specific measurable predictions before the experiment. Instead, they come up with vague predictions like "Black kids will do better on standardized testing if we implement teaching technique X" while completely failing to put a measurable number on what "better" means. How much will their test scores be improved, if the hypothesis is correct? It is rarely quantified before the experiment begins.

Once sociologists have established what political theory they want to prove, they conduct the experiment and mine the data only for research that corroborates their hypothesis. For example, if test scores are roughly the same, but kids finish the test faster, they will use that as evidence that they are doing "better." These post-hoc rationalizations are lazy science, and provide the wrong incentives to sociologists. No sociologist wants to admit that they wasted a ton of money and time on a hypothesis that turned out to be nonsense, or worse, that refutes their preconceived conclusion. This is especially true when it comes to ideas that they consider ideologically monstrous. When you are an activist social scientist who spend hundreds of thousands of dollars entering the field so that you can eliminate inequality, then the conclusive evidence that there is a massive IQ gap between African countries and European countries is something that you don't want to touch with a ten-foot pole. This is partially because the evidence offends you and flies in the face of what you **want** to believe, and partially because even if you were willing to show integrity and publish those results, none of the sociology journals would touch it, since they are likewise politically skewed heavily to the left. Simply by discussing an unpleasant truth, you would be destroying your own career. This is a problem because the IQ gap is real, and the damage caused by it is likewise real. No sociologist on the ideological left wants to talk about it because they would be subject to persecution from the same "woke" outrage mobs that murdered the founder of Cambrian Genomics. So instead they delicately tap dance around a major societal problem that they are all too cowardly to confront directly.

As a rationalist who is not afraid to tackle politically incorrect subjects, I believe that the IQ gap has been pretty conclusively proven by science. However, I believe that the data also shows pretty conclusively that the IQ gap is entirely caused by memetics, not genetics. This means that it is easily fixable simply by introducing the right cultural memes into Africa through advanced social engineering policies. In other words, children raised in a culture of success will tend to have better outcomes, regardless of whether they are black or white. This is empirically provable by the fact that America has already effectively conducted the world's largest scale social engineering experiment in transplanting black people from one culture into another, and the evidence shows that they do indeed have higher IQ scores as a result. Unfortunately, we can't fix problems when we are not even willing to acknowledge that they exist, or when society would stigmatize us and call us racist for pointing out self-evident truths. Avoiding unpleasant truths is not only bad science, but it leads to monstrous outcomes. Almost 1% of the population of Africa currently exists in a state of literal slavery. In my opinion, this could be easily solved through the right social engineering tools, but we are avoiding any real solution to the problem just because some people would be offended that we are insulting African governments. Look, we all live in the year 2020 so if your culture is so backwards that almost one percent of the people in your society are still slaves, I think you **deserve** to be insulted. Personally, I am offended by the mere existence of such massive incompetence, and I think that when a government is so corrupt and inept that their people are suffering needlessly, we have a moral obligation to step in and fix that problem by any means necessary, up to and including forcible replacement of that government. Reality is messy, and creating positive change often requires getting your hands a bit dirty.

Ultimately, I think that the reason that science has failed to advance in almost three decades is because scientists have an incentives problem. Like almost all incentives problems, this is yet another case of letting the foxes volunteer to guard the henhouse. Scientists are no different from any other Life: they do what benefits themselves most, while manufacturing justifications to rationalize their self-serving behavior. The fact that scientists tend to be smarter than laymen has no bearing on their ability to manufacture justifications, nor does it make them any less self-serving; it simply means that they do a better job of self-deluding themselves. A lot of human intellect is wasted on self-delusion, so the smarter a person is, the better they are at coming up with a persuasive heroic narrative to convince themselves that they are good people. Dumb people tell themselves dumb stories to justify the things that they do, and their narratives are easy to poke holes into. Clever people tell themselves clever stories to fabricate their post-hoc rationalizations, and so their stories tend to harder to debunk, because their narratives tend to be more logically consistent. Essentially, the reason science is broken is because scientists have no *skin in the game*. They are rewarded for exaggerating or distorting results, so they do that. If they are caught being massively incompetent, the punishment is generally inconsequential, so there is literally no incentive for scientists not to game the system. The easiest way to stop scientists from doing incompetent slipshod work is to enforce negative consequences on them when they show results that do not replicate. However, since a vast majority of scientists do shitty work, they know that any system which punishes incompetence will catch them in its net, and so they come up with all sorts of elaborate excuses to stop the system from being fixed. We need to accept that the crisis in science is largely the fault of our scientists, who seek to take credit for their own successes while avoiding accountability for their failures. We have the power to change this, but only if we stop deferring mindlessly to the dogma of self-styled "experts" and start forcing them to prove their expertise by backing it up with consistent results.

FIXING JOURNALISM

Over time at multiple jobs, I have observed an interesting phenomenon: that when you give stupid people power over smart people, it tends to have bad outcomes. There is no better example of this than the field of journalism. It has often been said in the past that a lie can travel around the world before the truth can even get its shoes on. But in the current age of viral clickbait and the 24-hour news cycle, today it is more true to say that a lie can get an angry mob to string up an innocent person on the nearest tree before the truth has even woken up and stumbled to its sock drawer.

Perhaps we should do something about that. During a recent political discussion over dinner, a close friend of mine told me "Nobody has the right to yell fire in a crowded theater." If we acknowledge the moral truth of this statement – that spreading lies can hurt people, and that our society has a moral imperative to punish people who do this - then I think we should obviously start with the media. In the current age of social media, these people have an enormous amount of ability to hurt innocent people by spreading inaccurate or biased information that has not been adequately fact-checked.

What would be the modern equivalent of shouting fire in a crowded movie theater? Some might say that it would be shouting "racist" or "sexist" in a crowded social media forum. In the past, being branded a racist or sexist carried little social stigma, but in today's society based around social media (where justice is crowdsourced, viral, and often fact-free) a person can easily lose their career, friends, family, and reputation, all on the accusation of a politically biased journalist. Based on the fact that social media systems are currently structured in a way designed to maximize outrage, I think it is perfectly clear that anybody publicly making a false accusation of some misbehavior is actively attempting to cause severe harm to the victim of the false accusation, which can obviously be equally as damaging (if not more) as shouting fire in a crowded theater. If it is reasonable to say that somebody ought to be punished for whipping up a lynch mob to harm a person in real life, is it not also reasonable to say that somebody ought to be punished for doing exactly the same thing online – especially when these false accusations can have equally damaging effects? As the kids like to say nowadays, "If you're throwing shade, bring receipts." Or as our president likes to say "WITCH HUNT!"

Perhaps a more personal example might help drive this point home. As you may have noticed, this book addresses some very controversial topics. This is necessary because evolution is not very politically correct, not does it conform to our current beliefs about right and wrong. However, understanding evolutionary psychology is very important because it allows us to mathematically predict the future using incentive topography as well as altering that future by setting up inflection points at certain key moments in order to change the course of history. This fact that Trump won – in defiance of all the so-called "expert" opinions - is proof that memetics and game theory works, that it is highly effective, and that sociology and economics are useless garbage. However, for the past twenty years, I have been unable to publish or express any of these scientific theories, because our social media model is currently structured to encourage the formation of lynch mobs against anybody who refuses to conform to our existing *historical narrative*. That means that if I had gone public earlier, I would have been called a bigot and a lot of outraged idiots would have quickly set up a petition to get me fired and ruin both my career and my social life. Since I don't believe in sacrificing my own well-being for a bunch of ungrateful idiots, I had to hold off on publishing this book for almost twenty years, waiting for precisely the right moment to strike – that perfect inflection point where I could prove conclusively and unquestionably that my *historical narrative* is accurate and the current *historical narrative* is outdated. How many tragedies could have prevented if we had had access to memetics and game theory twenty years earlier? How much environmental damage could have been avoided if climate change agreements had been written using game theory to optimize the incentives for everybody to do the right thing? When we consider history from a consequentialist viewpoint, the hands of modern journalists are covered in blood because there are so many tragedies that could have been prevented if we in the Intellectual Dark Web had been able to publish controversial science earlier. In their rush to condemn people for wrong-think, journalists have prevented the development of useful sciences for almost two decades and thus contributed to countless preventable deaths. Isn't it fair to punish them for it?

Of course, from a consequentialist perspective, fairness isn't important – the only thing that matters is the outcome. So let me rephrase this in consequentialist terms: the reason that journalists and media outlets engage in this evil and destructive behavior is because they are *incentivized* to do so. Articles that generate outrage generate more revenue for a publisher, regardless of whether the articles are accurate. Because journalists have substantially higher protections from slander accusations than average citizens, it is difficult to punish them for making false accusations. This means that journalists are incentivized to make false accusations and whip up lynch mobs, and have no disincentives to do so. In terms of incentive topography, we would say that the slope of the topography for journalism is angled towards sensationalist outrage and lies, because the incentives for doing so vastly exceed the disincentives. If journalists were incentivized to value accuracy over sensationalism – for example, if creating an online lynch mob with false accusations were to result in a public whipping – then their behavior would very quickly change, because journalists are only human and they respond to incentives just like the rest of us. We can change the field of journalism from being a deeply politicized profession occupied by evil liars, to an impartial field that is actually *useful* and valuable for humanity. All we as a society need to do is give journalists *skin in the game* by implementing laws that reward journalists for accuracy and diligence, and punish them for falsehood. This is not just limited to the practice of whipping up mob justice against innocent people; a lot of scientific journalism is also deeply inaccurate, and it results in vast social costs by stunting scientific and cultural progress. I think it would be nice to return to a period where we can actually believe what we read in the paper, and where journalists are incentivized to do thorough investigative reporting rather than just publishing wild gossip and speculation in their scramble to feed the insatiable appetite of media capitalism's 24-hour news cycle.

THE PATH WE COULD BE ON

A class I once took (melodramatically called "Speak like Obama") said that the end of any good communication should have a call to action. I feel inspired to try something like that here, and I hope you'll do me the courtesy of reading this section, even though there isn't very much *useful* practical information here. We may never have met in person, but I feel like in a way, we've gotten to know each other quite intimately over the course of this book. You've gotten an understanding of the way my mind operates, and in return you've given me a chance to shape your mind, because that is how information works. Knowledge is power, and when you digest knowledge, it also digests you, reshaping you into something better than you were before. So now that we share this special connection, I want to share something even more personal with you: my vision for the world. We've already spoken of how the world operates, and the tools that you can use to reshape the world. Now I'd like to share with you my personal beliefs about what we should reshape this world **into**. You may disagree with me about some of this stuff, and that's OK. No student ever agrees completely with their teacher. I suppose that if nothing else, I want this chapter to spark your imagination about the possibilities that are open to us. You are capable of so much and yet you settle for so little. I want you to strive for more.

Bottom Up Economic Stimuli instead of Top Down Economic Stimuli

When it comes to fixing structural problems, it's usually best to start with the low-hanging fruit – so let's begin with our economy. It has always been confusing to me why economic stimuli are driven from the top down rather than the bottom up.

What I mean by this is that economic stimuli (as currently designed) seem intended to offer cheaper loans to wealthy people in order to stimulate "entrepreneurship." In other words, presumably this wealthy caste of "entrepreneurs" will use that loan to create new businesses, which offer new products, and thus more money circulates in the economy due to people having the ability to buy those products. I'm not 100% convinced that the well-being of an economy can be truly reflected by the amount of money in circulation, but to be fair, I don't have decades of economics training. I'm fully prepared to argue with economists on the many topics where their dogma is obviously incorrect, but on the things that I'm slightly less knowledgeable about I'm prepared to give them the benefit of the doubt.

So my question is this: if the goal of economic stimuli is simply to circulate money, wouldn't the most efficient way be payouts (or tax cuts) to the poor? Studies show that when poor people receive windfall income, they tend to spend it exceptionally fast – much faster than rich people. In other words, if you give a dollar in stimuli to a wealthy person, they will tend to park it in an ETF or stock, investing it so that they can make more money in the long term. This is good for the wealthy person, but terrible for the economy, because the goal is to circulate cash. When you park money into long-term investments, that money has only circulated once. Some might argue that banks stimulate the economy by offering loans that supposedly get used to start new businesses, but in my opinion that kind of thinking is old and outdated. Everybody knows that nowadays banks make a lot more money through transaction fees and other services.

By contrast, if you give a dollar in stimuli to a poor person, he goes to a convenience store to buy cigarettes, or booze. The convenience store worker then buys a metro transit card. This is used to pay a government employee who might then use the extra dollar to subscribe to HBO, and so forth. While I do think it is tragic that poor people tend to spend windfall money more rapidly rather than investing it, I'm not judging them. My point is simply that we know that money from subsidies circulates much more when it is given to the poor rather than the rich, and multiple studies have already proved it. Trickle-down economics is a lie; the truth is that trickle-up economics is far more beneficial to a country's GDP. It's hard for budding entrepreneurs to open up businesses catering to the mega-rich, partially because the rich don't part with their money easily and partially because it's often difficult for an aspiring entrepreneur to gain access to them in order to deliver your sales pitch. The poor, on the other hand, are practically giving their money away, which is the perfect environment for start-ups to blossom. Eventually we will need to change this cultural habit in order to eliminate poverty, but in the meanwhile, why not take advantage of it to stimulate our economy? As scientists, we need to interact with the world as it **is**, not the world as we **wish** it was.

This may sound cynical, but I believe that the reason most of our economic stimuli is based around economic subsidies to the rich rather than to the poor is because the rich have better lobbyists, so they convince the government to do what is in their self-interest and then use *motivated reasoning* to rationalize their behavior by convincing themselves that it is the correct course of action. As long as I'm being cynical, I could also point out that if we managed to test out my hypothesis about trickle-up economics and prove that it works, then it could be a perfect tool for things like trade wars – stimulating the economy while gaining the public support of the masses at the moment that a leader needs it most. This is what I mean when I say that it is so critical for societies to accept game theory and memetics as legitimate sciences that deserve further testing and study. The methodologies that we have discussed in Chapter 4 are crude and unsophisticated tools compared to the marvels of system-building that are possible.

Reparations and an End to Racism

Considering that I view imperialism as a good thing which provokes important cultural evolution, it may surprise you that I am in favor of reparations for black Americans. This may seem especially strange because I strongly oppose identity politics, which I view as a gateway to demonizing successful white people (in much the same way as successful Jews were demonized prior to World War 2). I would like a chance to justify my views, or at least explain the logic behind them.

Democrats tend to believe (or at least **signal** that they believe) that the problems in the black community are entirely the fault of white people. From this, they derive the conclusion that white people should feel guilty about slavery and try to make it up to black people with reparations. I think that this guilt-driven perspective is nonsense. First of all, black people would definitely have enslaved white people if the situation had been reversed, and in fact history shows that whenever black people had a technological or cultural advantage, they **did** enslave white people. So feeling guilty about slavery is like a tiger feeling guilty that it has sharp claws. When you live in an environment where the only choices available are being predator or prey, then obviously it's far more desirable to choose the role of predator. Preying on those who would otherwise prey upon you is not a sin; it's the way of the universe. Strength is nothing to be ashamed of, especially since evolution has a tendency to kill weak species, and it would be very unusual if humanity was an exception to that rule. Personally, I hope that as humanity evolves, we become stronger and more resilient: because as the complexity of our society grows, there are increasingly fewer challenges that can be resolved by "hugging it out."

Republicans tend to believe (or at least signal that they believe) that racism is no longer an issue, based on the fact that other minorities that immigrated to America have had more success than black people. From this, they derive the conclusion that the problems within the black community are entirely the result of their own defective culture, and that black people need to take responsibility for this and "pull themselves up by their own bootstraps." I agree that black culture could be improved, but I think that this analysis totally overlooks the economic factor. It is easy to have good cultural and parenting habits when you're making $100k a year on a 40-hour workweek. But when you're stressed out from hustling two jobs to feed your family, making $50k a year with a 60-hour workweek, it is much harder to find the time and energy to instill good cultural habits in your children. I imagine it's hard to even get enough **sleep** in that kind of environment. It's also very difficult to develop good spending habits and save money when your workplace forces you to spend thousands of dollars to alter your natural hair simply in order to keep your job. So this attitude towards the black community seems to me like it comes from a position of great financial privilege.

This leads me to a point that I think both sides have completely overlooked: the fact that racism is downstream of economics. It may surprise you to learn that when the Irish first immigrated to America, they were not considered white and were treated with the same kind of contempt and discrimination that black people have experienced. Then they became financially successful, and boom – the racism against them magically disappeared. Suddenly they were white, and in hindsight it seemed silly to imagine a time that they had **not** been white. The same pattern would later repeat itself with the Italians, who were considered a dark-skinned minority when they first arrived in the U.S. Once they got money, boom – they were accepted as white, and moreover they had **always** been white, since people changed their views retroactively to accommodate the new *historical narrative*. Do you get what I'm saying? Money changes the way that people perceive color, because ultimately the only color that really matters in our society is the color of money. Based on their financial success, I predict that Asians will be the next group to be assimilated into the circle of whiteness as it expands. In fifty years, our descendants will be saying things like "Wait, there was once a time when people thought Asians were not white? What a bunch of bigoted idiots!" Before you laugh dismissively at this idea, perhaps you can explain why a lot more white supremacists are suddenly dating Asian women? Race is a flexible concept, even to white supremacists.

Another way to frame this is that all of the conversations that we are having about oppression and race are totally nonsensical. We should actually be having a conversation about money. Once we solve the money issue, race will become irrelevant. The trouble is that neither Democrats nor Republicans genuinely **want** to solve this problem. Democrats don't want to solve it because they have a nice scam going in the form of a grievance-industrial complex whose entire purpose is to make African-Americans and other minorities feel oppressed and resentful against white people. This allows Democrats to create all sorts of wasteful "diversity programs" that are supposedly about elevating minorities from poverty, but actually serve as more of a platform for liberal HR staff in corporations and schools to indoctrinate employees and students into woke political propaganda. Meanwhile, Democrat politicians play the race card to advance their careers, while Democrat entertainers use their minority status to advance their own careers by calling for more minority "representation" - which is really shorthand for "fast-track my career in this industry or I'll publicly call you a bigot." Now maybe my galaxy brain is just too vast to comprehend the small nuances of this discussion, but I genuinely don't see how it benefits the black race as a whole for Beyoncé to sell a few extra T-shirts at her concerts. It seems to me that what black people really need is money, not "representation," and this whole diversity industry with its quotas and affirmative action and "representation" is just a way to distract black people from the fact that they aren't getting actual **money**. The diversity industry will **never** pull black people out of poverty, because if it did, then the industry itself would stop existing, and a lot of well-paid liberal morons who have no actual job skills beyond "professional wokeness" would be out of a career.

The reason Republicans don't want to solve the problem is more straightforward. Unlike the Democrats, who use *motivated reasoning* to rationalize their own behavior in a way that lets them affirm a delusional *heroic narrative*, Republicans already know that this is about money and they just don't want to give it up. It's flat out greed, with maybe a small side order of racism. It may sound like I'm being far more critical of Republicans than Democrats here, but that's not my intent. Frankly, I have a certain respect for people who are honest with themselves about their own motivations, even if those motivations suck. Republicans use the "defective black culture" argument to push the blame for black failure upon black society, handwaving away the highly relevant fact that the economics of the situation create a self-perpetuating cycle that makes it very hard for black people to fix their own culture.

My perspective is entirely different. As a consequentialist, I don't particularly care whether the blame for black poverty falls more on black people or white people. This finger-pointing is counterproductive and useless. I would simply like to live in a stable society that doesn't collapse into violent anarchy. Regardless of how we got to this point, black people are part of our society now, and I think that much like the Roman empire, our society will be more stable if we can get them to identify as Americans first and their ethnicity second. To instill this sense of national pride into black society, we need to incentivize them by making them stakeholders in our success, ensuring that the fortunes of black people rise and fall in conjunction with America's own prosperity. This gives them *skin in the game*. Most data that I have seen suggests that at the current rate of progress, it would take anywhere between two and three **centuries** for black people to achieve the same level of economic prosperity that white people have reached. That is the worst incentive ever for black people to develop a sense of national pride. If I was told that people of my race would have to patiently be an underclass in America for two or three centuries before achieving economic parity, my reaction would be somewhere along the lines of "Fuck America, and fuck each and every one of you! I hope this goddamn country of racist fuckers gets hit by a meteorite." I really can't blame black people for being a bit unhappy with the current rate of progress. I'd be a lot more upset in their place.

Of course, giving black people money is only half the battle. Studies show that black people are very bad at keeping money within the black community, because they have not yet developed a culture of generational wealth. In simpler terms, it's hard to develop cultural norms about saving and investing money when your people don't have that much money to practice these norms with. If we gave black people a single lump sum money transfer, then frankly, they would probably blow it on short-term status/luxury goods rather than making wise investments. Fortunately, culture is one of the easiest things in the world to change, with the right training methodology and effective teachers. For example, I'm changing your culture right now, just by writing this down for you to read. It doesn't even matter whether you want your culture changed, because once you learn something, you can't unlearn it. When you digest knowledge, it also digests you.

To make a long story short, my plan to fix racism is to establish schools in predominantly black neighborhoods which teach children about financial responsibility and strategic investing, with maybe a sprinkling of the Dark Arts thrown in. The schools will not be **exclusively** black (because segregation encourages racism) but due to their locations in poor black communities, the majority of the students will be black simply to reflect the ethnic diversity of the surrounding neighborhood. Simultaneously, we institute reparations in the form of recurring structured payments that gradually diminish over time, so that black society will not have to deal with sudden economic shocks that they are unprepared to navigate wisely. Affirmative action programs will be repealed at the same rate of change, so that black people can't double-dip from economic equality programs in a way that is unfair to other races. Also, the fact that affirmative action programs will be unwound slowly and in measured steps will give the professional grievance-studies majors who run these diversity programs opportunities to retrain and learn some actual job skills that extend beyond pushing their morally bankrupt ideologies onto everybody else. What all of these programs have in common is that they target the economic factors that fuel racism. Once black people are financially prosperous, the circle of respectability will naturally expand to include them.

I mentioned earlier that I believe that trickle up economic stimuli work more effectively than trickle down economic stimuli, and I think that reparations would be a perfect test of this hypothesis, since black people are one of the poorest groups in America. Whenever a structural change is implemented in any large system (whether by design or by accident) it is always important for Dark Rationalists to make predictions, record as much data as possible, and then compare hypotheses to results. This iterative process improves the science of Game Theory substantially, allowing increasingly potent results. From a certain perspective, some might say that society is a laboratory, and we are its researchers.

Advanced Robotic Workforce and Universal Basic Income

Let me ask you a question that might sound stupid at first: why do we work? As far as I can tell, there are two different schools of thought on this question.

The first school of thought is that we work to contribute to society, build something positive, and feel like we are making a genuine difference in the world. This school of thought is belied by the fact that 99.9 percent of all jobs do none of those things.

The second school of thought is that we work to make money, so that we can afford food and healthcare and avoid starving to death in the street or dying of an easily treatable illness. This school of thought is harder to refute, but if you subscribe to it then you may find it alarming that most analysts estimate that half of all the jobs in existence today can easily be replaced by machines, based on our current levels of AI and robotics technology. That means that we have three options: we can either avoid developing AI (and eventually get conquered by another society that does not have such restrictions), we can resign ourselves to mass death from starvation and illness (or from a revolutionary war by hordes of poor people who are not onboard with the whole "crawl into the gutter and die" program), or we can make some radical changes in the way that our society operates. Personally, I am in favor of the latter option.

Our society has always walked a dangerous fine line between socialism and capitalism. There's an old joke that under a socialist system, you find out how lazy and evil people are, and under a capitalist system, you find out how **industriously** evil people are. I think that joke nicely summarizes some of the problems that we're going to see with UBI and industrial automation. Industrial automation has the potential to create vast profit for society, and it's going to be a difficult question figuring out how to divide that profit. If we allow entrepreneurs and venture capitalists to reap **all** the rewards of a fully automated workforce, then we end up in a dystopian future of libertarian tech-nobility lording it over hordes of poor serfs. But if we socialize **all** of the profits to the public, then nobody has any incentive to automate (or innovate) and eventually our whole society collapses. We need to strike a balance of incentives: allowing smart people to become very rich if they come up with ideas that benefit society, but taxing the wealthy enough that their heirs don't end up living comfortably on vast amounts of generational wealth just because their great-great-grandfather had a good idea. I think that innovative people are entitled to keep what they earn, and pass some of it down to their kids. But I don't think that their kids deserve **as much** wealth as the original innovators had (unless they turn out to be just as talented in their own right), and I think that each successive generation of descendants has less and less of a legitimate claim to their ancestor's wealth. Because wealthy people have disproportionate influence in our society, a gradual reduction of generational wealth ensures that our leaders will always have a certain degree of Darwinian evolutionary fitness, rather than being inbred idiots whose main skill is socializing. If the rich have no fear of being toppled from their thrones, they inevitably become just like the French nobility that preceded them – arrogant, ignorant, and complacent. When this happens, they need to be treated just like their forebears and introduced to a guillotine. The world is in crisis mode right now and we have no time for leaders who are not up to the job of actual leadership.

The difficulty with UBI will be finding that sweet spot in the incentive topography – the place where the owners of capital are taxed a low enough amount that they are still incentivized to innovate and automate, but high enough that we never get families of generational wealth acquiring a stranglehold on power simply by virtue of inheriting money. I think that this will take a combination of finesse and political astuteness, but I also think that we have the right balance of skills to make such a dream successful. At first it will start with a gradual shortening of the work week, but in the far-distant future I hope for a time when nobody has to do any work unless they find it intellectually or emotionally rewarding.

The main difficulty with UBI has been selling it successfully to the public. Every time somebody brings it up, a lot of wealthy people start talking about how it's "unrealistic" or "economically unfeasible." Personally, I believe that the best way to sell people on an idea is to demonstrate proof of concept, since it's very hard to argue that an idea is unrealistic when you've already implemented it successfully. Do you remember how I mentioned earlier in this chapter that bottom up economic stimuli tend to have more effect than top down economic stimuli? And how I also mentioned being in favor of reparations for American blacks who are the descendants of slaves? To me, these things are very closely tied to UBI and industrial automation, because everything we do has ripple effects. When we first implement reparations, it's very unlikely that we'll do it as a one-time payment. First of all, that would bankrupt the United States, and second of all, it would do black people a huge disservice to give them that amount of money all at once without a corresponding amount of training and education in responsible financial practices. Due to the incentive structure of capitalism, such a course of action would end up with entire industries whose main focus would be conning black people out of their newfound wealth.

In other words, the only realistic way to implement reparations is through a structured series of payments that gradually diminish over time. By a complete coincidence, this bears a striking resemblance to the way that UBI is supposed to work. On that note, let me give you a hypothetical scenario. Suppose that we implement reparations and we have a bunch of Dark Rationalists in a think tank somewhere, measuring the effects that such a bottom-up stimulus has on the economy. And suppose it turns out to have a really beneficial impact on the economy. Maybe it even helps us win some hypothetical trade wars with a few nations who thought that they could challenge our primacy. It seems to me that in that situation we would be perfectly positioned to prove conclusively to the public that UBI could be economically viable. You see, this is how a think tank full of Dark Rationalists operates. We make hypotheses, then we test those hypotheses, collect the data, use said data to refine or revise the hypotheses, rinse and repeat. When an experiment is successful, we see if the effects are reproducible on a larger scale, and we continue this pattern until the world changes from what it is to what it should be.

But who knows? Maybe I'm wrong about UBI. I'm just one lonely game theorist who currently has no financial backing for my research, so social experiments that span entire societies are new for me. It's possible that we'll implement reparations, do the math afterwards, and it'll turn out that the bottom-up economic stimulus wasn't quite successful enough to justify the huge costs that it would take to implement UBI for everyone. But here's the important question: will we regret having done the experiment? The worst-case scenario is that we gain some valuable data that helps us adjust course to better steer the direction of society, and the best-case scenario is that we create a literal utopia. I think that this is a course of action that is at least worth investigating.

Religions of Mindfulness

I think that people gravitate towards religion because on some level, we all want to be a part of something bigger than ourselves. Our modern narrative tells us that there's nothing more to life than the physical realm that science tells us exists, but I hope I've already demonstrated several times in this book that at least two scientific fields currently in widespread use are nothing more than politicized nonsense. So maybe there are other scientific narratives that we should be questioning as well.

To be clear, I'm not saying that I'm a particularly religious person. I'm just saying that we should be open to alternative narratives, and religion (if done appropriately) is certainly one of those. The postmodern neoliberal narrative that we currently live in seems to force us into valuing ourselves based on our consumption. Even our current economic models are all based on the assumption of continual growth, without any consideration of the fact that growth cannot last forever - or that our use of the current economic model makes it inevitable that once growth stops, a painful crash is soon to follow. But it's not just economics; this consumption mindset seems ingrained in every part of our society. Our environment is the same way – we use up more and more resources, thinking that our current way of life or population growth is infinitely sustainable. What will a crash look like in that situation? It's as if the entire human race is nothing more than a horde of locusts whose sole purpose is to multiply, strip the planet bare, and then die off in a mass extinction. And if anybody points out that maybe this entire philosophy is morally bankrupt and alarming, our elites – steeped in a culture of wealth and privilege – scoff and tell us to be "more realistic." Excuse me? I think that I'm the **only** one being realistic here. Our elites have no *skin in the game*. With their wealth and position, they can insulate themselves from the negative consequences of their own shortsighted policies. In other words, when the food riots start, you can bet that our global leaders will still be eating well.

In a world that is on the verge of a massive die-off, this seems like the perfect opportunity for religion to step in. Our environmental problem is a Defector's Dilemma, and such problems require group organization and coordination. We live in an atomized and materialistic world which is gradually stripping us of our privacy, our dignity, our environment, and eventually our lives. This seems like a void that is begging to be filled by a religion that actually uses its power to actively solve the problems which are afflicting each and every one of us. I mean, religion was practically **designed** to solve group coordination problems. Back in historical times, it used to be that when the pope said "Hey, look at those morally bankrupt people over there," people would grab their weapons and start slaughtering millions of people, based on nothing more than the say-so of a single dude with a funny hat. Granted, in historical times this power was generally abused, because human beings are easily corruptible and something about sitting on a throne and wearing lots of bling seems to bring out the worst in people. But imagine if that power were used for good, to solve actual problems like complete environmental catastrophe and mass human extinction? I think it could be a very *useful* tool. In fact, I imagine that even just the threat of such drastic religious mobilization would provide a lot of leverage in terms of getting your average politician to re-evaluate their priorities.

And hey, here's an interesting thought: what if in addition to fixing the problems that currently plague the world, our hypothetical religion actually helped fix people's problems on a personal level? You know... like religion originally **used** to do? Most of our last decade of sociology and psychology research is complete garbage due to the way that those scientific disciplines have been politicized, so if we had a religion that actually used **real** psychology (as well as the Dark Arts of Rationality) to help people solve their problems, I think it would have a real first-mover advantage. I think it might even become so popular that it could drive our current generation of psychologists out of business. It's something worth thinking about.

Environmental treaties that actually work

We generally don't like to talk about this in public, but fear and the self-preservation instinct play an important role in group compliance. So here's the bad news that many naïve optimists are going to have a hard time accepting: we can't even begin to solve the environmental crisis without making examples of a few defectors who refuse to play ball. Our current environmental treaties don't work because they have no teeth – they are primarily nothing more than wealth transfers from developed nations to undeveloped nations, with the undeveloped nations promising that they will use the funds in environmentally and economically beneficial ways. However, instead of spending the development funds as promised, the politicians in these developing nations turn around and embezzle most of the money. They don't care about the well-being of their people, because they are not **incentivized** to care. They don't even care about being honest, because the developing nations that provide them with this money don't **punish** them for dishonesty. Naïve politicians like Barack Obama cheerfully offer handouts to corrupt criminal regimes, smiling as these dishonest and corrupt glad-handlers offer up bullshit statements about how they will do the right thing. Meanwhile, there are districts in the southern U.S. without access to adequate sewage systems. Why are we giving billions of dollars to corrupt regimes so that their government officials can buy superyachts while our own citizens are dying from lack of adequate infrastructure? Is this how environmentalism is supposed to work? Also, I can't help noticing that most of the Americans impacted by this inadequate infrastructure are black. Isn't racial inequality something that the Democratic party is supposedly really concerned about? I mention this only because I see a direct correlation between naïve presidents giving away massive bailouts to the wealthiest 1% - both at home and abroad – while disenfranchised black Americans lack money for the most basic necessities of life. We live in a world where money is life, and bad fiscal policy literally kills people. That is why I consider stupidity a sin, especially when it comes to public office. It doesn't matter what your **intentions** were; the only thing that matters is the **outcome**. It means nothing to talk about improving life for the poor when your dumb policy decisions do the opposite. As the saying goes, the road to hell is paved with good intentions.

One of the major problems that I see in our environmental treaties is that a lot of the points I've made above are not taken into account. An agreement that cannot be enforced by inflicting punishment on anybody who violates it is a pointless agreement. In fact, it's **worse** than pointless, because by transferring resources to people who are failing to live up to their agreements, you are incentivizing them to **continue** with that behavior. I've pointed out earlier in this book how first world countries with good intentions incentivize a lot of bad behavior among third world countries by giving them development funds with no strings attached, and environmental treaties that lack teeth are yet another example of that.

However, in my opinion, a more significant problem in terms of environmental progress are the environmental activists. The sad fact is that most climate change denialists come from within the environmentalist community themselves. I suppose I should clarify what I mean by "denialist," because on the face of it, this may not make sense to you. When I talk about a denialist, I am referring to somebody who sees hard evidence of something that they don't like, but refuses to believe it because the truth is unpleasant to accept.

Here is the truth that the climate change denialists in the environmental activism camp don't want to accept: we are already too late. All the evidence indicates that we have already passed the tipping point, which means that right now, there is absolutely nothing we can do to prevent a mass die-off. It is only a question of whether this mass die-off kills 30% of humanity, 50%, or whether we are going to go for some **real** high scores like 80% or 90%. In short, the time has come to accept that this trend has some real momentum, and we need to change our efforts from preventing climate change to finding ways to mitigate it. Unfortunately, environmental activists are the biggest obstacle to this goal, for a number of reasons.

1) Often, they are financially vested in the system of preventing climate change. Shifting resources away from organizations that work to prevent climate change

and moving them towards organizations that work towards climate change mitigation would eliminate many of their jobs and financial resources.
2) It would force them to accept the fact that the cause that they have invested their lives into has been an utter failure – and consequently, that they themselves are utter failures. More significantly, the public would see them for the failures that they are, which would be a significant loss of status for them. When it comes to status games, most climate change activists are no less selfish than anybody else in the world. They may be playing a different status game than most other people, but they hate losing just as much.
3) Accepting that not everybody will survive the coming climate apocalypse means that humanity will need to shift to an entirely different value system, in which it is OK to let groups die if they aren't pulling their weight to sustain civilizational advancement. Under the faulty historical narrative that most environmentalists subscribe to, everybody can still be saved if we all work together. This is why many passionate environmentalists – such as Alexandra Ocasio-Cortez – are also advocates of open borders. They believe that America has the responsibility and the duty to take in climate-change refugees from the rest of the world. The truth, however, is that America and a few other countries are lifeboats for the climate-change apocalypse, and like all lifeboats, they have limited seating capacity. If you try to fit too many people into a lifeboat, the whole thing sinks. The same thing happens to civilizations burdened by resource scarcity with too many mouths to feed. As Africa slowly dies due to Malthusian pressures, this will become increasingly obvious. At that point, the first world nations of the world will have to make a choice – do we want to take in the refugees and increase our own nation's risk of complete destruction, knowing that each additional refugee we take in means that one of our own citizens may have to die? Or do we want to simply seal off our borders with weaponized AI and let the less advanced countries of the world deal with the unfortunate consequences of their own failures to plan ahead and government choices?

The significance of this third point cannot be understated. I mentioned earlier that the system shapes our personalities and cultures. In a world where survival is easy, people become generous and compassionate. In a world where survival is hard, people become ruthless and calculating. In other words, we are talking about a complete cultural shift in our values. Right now, we live in a world of abundant resources, so we are taught that every life is sacred. But in a world of scarcity, our belief in the sanctity of human life goes right out the window. Nobody is assumed to have an inherent right to exist - instead, their right to exist is based entirely on whether they contribute to society. This will particularly impact the idle rich, since they are resource-heavy targets that don't contribute much to society. Furthermore, their hypocritical behavior about climate change is something that a lot of people will remember. In other words, once the grim nature of our situation becomes apparent, a lot of knives are going to start getting sharpened, and the people who are perceived to have been contributors to this problem are going to have to start fearing for their lives. I'm not just talking about Republicans who refused to believe in climate change – I'm talking about Democrats who advocated for open borders and globalism. Our only chance at survival is to reduce population, develop AI, and build supply chain independence so that when the other nations of the world start collapsing, we do not collapse with them. People who are not onboard with this plan are contributing to the problem, and I fully expect that they will be called to account for their deficiencies later - when the full extent of this crisis becomes clear and we as a society are deciding which of our elites carry the blame for allowing this to happen.

This may sound like a cruel, perhaps even sociopathic, point of view. But we have countless amounts of scientific data to back it up. We know without a doubt that humanity has passed the environmental tipping point. There are reams of climate change reports that previous administrations have financed (because apparently "raising awareness" – ie, deliberately keeping this issue alive to get votes - was more important to our esteemed colleagues than actually taking action to **solve** the problem). We also know without a doubt how large crowds behave under conditions of resource scarcity, because there are countless historical examples to draw data from. The only reason that my description may sound unreasonable is because you don't **want** to believe it. In much the same way that the citizens of the Roman Empire refused to acknowledge that the Empire was in a state of collapse, even when the barbarians were right at the gate, our elites don't want to acknowledge that the existing world order has failed the citizenry. I suppose you could say that our existing elites are the Valerians of the world, and I am simply following the wise example of Marcus. I have served loyally for most of my life, but now that the truth about global civilizational collapse is so self-evident, I refuse to continue following incompetent leadership into the abyss created by their own bad choices. We are operating under a faulty historical narrative, and it is time to update our assumptions and acknowledge the reality of our situation.

To solve the problems of climate change, we first need to accept the reality of the situation. The reality is that we have already passed the tipping point at which everybody on the planet will survive, so we need to stop living in denial of that reality and ask ourselves some hard questions about which societies *deserve* to survive. This is a team effort that requires cooperation, unity, and integrity in dealing with one's allies, and so I think that the civilizations that deserve to survive are the ones that are optimized for it – in other words, the civilizations that best exemplify those values. That means that we need to do four things:

1) Set up an organized Environmental Coalition of the nations most committed to solving climate change. This organization of elected officials would enact laws by simple 51% majority (thus solving the *heckler's veto* problem) and would be empowered to ferociously punish any member nations that did not abide by its resolutions (thus solving the *defector's dilemma*).
2) Publicly acknowledge that the only purpose of this organization is to focus on the survival of the nations **that are members**. Any nation that joins the coalition and abides by its rules gets representation and assistance. However, any nation that chooses not to join the coalition does not receive any help or financial aid whatsoever. Even if a whole country is dying and there is a massive refugee crisis, coalition nations won't help anybody outside of the coalition – they just close the borders and let everyone in that country die. The vast discrepancy between the benefit of joining the coalition (your nation has a much higher chance of surviving,

thanks to the help of your allies) versus the punishment of not being part of the coalition (your nation is on its own in the face of global warming, meaning that you're probably going to die) maximizes the slope of the system's *incentive topography*, making it increasingly probable that as time goes on and entire nations start dying, even the proudest and most arrogant government leaders will see the value of bending their knee and accepting the authority of the Coalition. It also helps solve the *defector dilemma* because defectors can no longer freeload on the effort of team players.

3) Actively support this Coalition with worldwide memetic propaganda, in both member nations of the Coalition and non-member nations. The goal is to utilize *Teach's principle of preference cascades* to create a preference cascade effect. Because citizens that are not part of the coalition will not be allowed to emigrate to coalition nations (instead, they must simply deal with the repercussions of their own government officials' poor choices), then the average citizen suddenly has a lot of *skin in the game*. They cannot afford to elect corrupt or inept government officials who make poor choices about environmental resilience, and as their country's situation worsens, they will increasingly pressure their government officials (whether with votes or bullets - depending on whether they live under a democracy or an authoritarian system) to join the Coalition.

4) Utilize game theory to maximize the success of Coalition planning. For example, at the current moment in time, pollution is largely driven by population, due to the rapid spread of industrialization. That means that we need to reduce population growth rapidly. Instead of using brute-force policies that don't work (like China's failed one-child policy) Coalition nations would be forced to use more effective and scientifically proven game-theory methods of birth control, such as improving women's access to economic opportunities, education, contraception, and the ability to financially control their own lives. Additionally, any economists coherent enough to have their careers survive the shift to the game-theory paradigm would be forced to update their economic models away from the unsustainable growth model that currently guides our policy makers.

5) Leverage AI in both industrial automation and the military to supplement societal advancement whenever possible, so that a nation's ability to project power is no longer based on its population size. Additionally, once AI robots are realistic enough to no longer generate an uncanny valley effect, AI sex companions could be further used to reduce population growth as well as mitigating aggressive tendencies among the male population, since it is another easily proven fact that societies with lots of single men tend to be more violent than societies where men are getting laid regularly. This means that in a world where men can easily get an artificial companion made to their exact physical and behavioral specifications, violence and population growth are both substantially mitigated, because the only people who would ever reproduce would be those self-aware enough to value a genuine human connection over digitally perfect artifice.

This may seem like a very strict and desperate plan of action, but we need to be honest with ourselves – we are in desperate times. Mass migration from climate change has already started (way ahead of forecasts), and a wave of refugee crises is traditionally the first in the series of dominos that has historically led advanced societies to crumble. Right now, politicians and their constituencies are not desperate enough to enact all of the changes that I have suggested, but as our environmental crisis gets worse and our current approach of generosity and compassion fails to solve the problem, the remaining societies which survive will become increasingly ruthless and calculating. When we eventually reach a cultural tipping point where people are willing to do literally anything to survive and preserve their civilization (in other words, a point where the outdated *historical narrative* shifts to something more *useful*), this list of recommendations will hopefully serve as a *useful* blueprint for future leaders to solve the problem of climate change much more effectively.

One world, One government

There are some goals that can be accomplished in a single person's lifetime, and there are some goals that can't be. I believe that a one world government is one of the latter goals. In an era of thermonuclear weapons, there is no realistic way to forcibly unite the entire world through military conquest: we have missed the boat on this. That means that more subtle levers need to be used. Fortunately, subtle social control is what Dark Rationalism excels in. When we enhance current Game Theory and Memetics models with more precise AI-generated algorithms that allow us to calculate exactly how people will react to various memetic inputs, we can initiate global change simply through good advertising. The wars of the future won't be fought with missiles or tanks; they will be fought with AI-generated propaganda and manipulation of other people's perspectives. Some might even say that the first skirmishes have already begun. Already, we have AI-generated "propaganda superweapons" that are considered too dangerous to be released to the general public.

In some ways, conflict is inherent to Life. As long as Life exists, it will do whatever is necessary to increase the chances of replicating its own pattern, typically by acquiring resources and power. However, I think it would be in everybody's best interests to ensure that most conflict is social rather than military, and ultimately this can only be achieved by creating a single global government. As long as multiple governments exist, there is always the chance of war breaking out, and the proliferation of weapons of mass destruction means that any sort of military escalation can always spiral out of control, resulting in the destruction of our entire species. While the chance of that happening is very small on a year-to-year basis, when viewed over a massive timescale the probability of an apocalyptic war is almost guaranteed, especially if we continue along our current path of cultural evolution. In the long-term, if we want our species to survive indefinitely, we need to put an end to war by ensuring that the entire planet is united under the auspices of a single government. Preferably, this would be a benevolent government that allows every citizen to live up to their maximum potential.

How would we do this? Some would say that uniting the entire planet is an undertaking of such scope that it is effectively impossible. Then again, the people who say this are typically the same people who believe that predicting and manipulating the future is impossible. If we allow ourselves to be shackled by the intellectual constraints of people of limited imagination, then obviously we will never achieve anything of value. It is only when we peek behind the Great Lie and open our eyes to the full scope of what is possible that we gain the ability to shape the future in a way that is desirable to us. But that is a job for future generations. My goal is simply to give you the tools necessary and demonstrate proof of concept, while helping to eradicate the backwards superstitions and deluded beliefs that currently pass for science. If that phrase sounds too abstract and spiritual, perhaps a more pragmatic frame might give more explicit clarity: I am showing you the tools and techniques of the Digital Age with the expectation that you **use** them, experiment with them, and improve upon them. The reason our world ecology is in a desperate state is because of humanity's greedy and selfish behavior. Previously, difficulties in getting large groups of people to act together in a coordinated manner made it tricky to change unhealthy societal behaviors. When you diligently practice the techniques of Memetics and Game Theory, eventually you will find that getting large groups of people coordinated is actually quite easy, which greatly facilitates the process of changing societal norms. This means that you have all the tools that you need to unify the world already at your disposal; you simply need to polish them up a bit, modify them slightly to fit your specific situation, and then deploy them.

The Science of Progress

I hate to revisit this topic again and again, but it really does bear repeating: we have a huge problem with our process of scientific discovery. While economics and sociology are the fields that have been most badly infected by a corruption of free inquiry, the rot is spreading. This is very important because our planetary ecosystem is dying due to climate change and overpopulation, which means that we are operating on a very short timetable. If we do not manage to make enough scientific progress to reverse climate change by a certain point in time, it is basically game over for the entire human species. I think that this is a bit problematic, as the young folks say.

In broad terms, the reason that science is rotting away at the core is because we defer to scientific "experts" who have no genuine expertise. The tests of replicability and predictability are critical to any scientific experiment. Without incorporating those tests into the scientific process, we have no way to determine whether any new scientific theory is legitimate or totally fake. What makes this such a problem is that science is an iterative process. Imagine science as a temple being collaboratively built. Every new theorem is a brick that is laid on top of another principle that was set down previously and accepted into the "scientific canon" - in other words, our *historical narrative* of how the world works. When any aspect of our scientific canon is incorrect, then every theory and principle that derives from it will also be flawed. Faulty scientific theorems that are accepted as fact are very harmful to the process of building our temple of science – they are like hollow bricks that eventually crumble under pressure, causing massive damage to the entire edifice as everything that rests on them comes crumbling down. In our current scientific process, we do not devote enough time to testing the structural stability of each brick – we assume that just because a theory **looks** sound, that it is worthy of being published and accepted as scientific fact. When this assumption is incorrect, it results in entire areas of science built on nothing but hot air and pompous nonsense. This is how we get fields like astrology, alchemy, phrenology, and economics – entire edifices of nonsense built on flawed assumptions which gradually accreted over time until they got treated with the same status and respect as **real** science. In the present day, we even use these nonsense fields to make financial and policy decisions. Eventually these scientific fields end up collapsing when they come in conflict with reality, but it takes a lot of time and effort to debunk this nonsense because by that point in time, a lot of elites have significant prestige invested in the existing status quo. This means that the elites push back quite vigorously against the idea that their narrative is wrong, since it would invalidate their expert status.

      This elite pushback is the reason why the only effective way to get rid of an entrenched pseudoscience is to weaponize **real** science to eliminate the power base of the elites. For example, take the 2016 election. Do you think Hillary or Obama would have ever admitted that all the "great advice" that they were getting from sociologists or economists to set their policy was total nonsense? That Obama totally screwed up the way he handled the housing crisis and Title IX? Of course not, because then the public might ask legitimate questions about why the advice of charlatans was being used to set public policy in the first place. If I had somehow managed to get an audience with Hillary Clinton (back when she was a serious political candidate instead of an embarrassment) and told her "Yo, all this economics and sociology stuff that your advisors were telling you about is pure bullshit – you should listen to me instead," what do you think would have happened? You don't need to be an expert in Game Theory to figure out that she would have laughed at me. Well, who's laughing now? Democrats are finally starting to take the "unconventional" sciences of game theory and memetics a bit more seriously. Democrats would never have reached this degree of self-awareness if their "surefire win" hadn't lost so thoroughly to an unconventional campaign by Donald Trump. As the old saying goes, very few people appreciate being publicly bitch slapped, but it is an undeniably effective way to immediately grab their attention.

      Regardless of the final outcome, the fact that it took a humiliating presidential defeat to make Democrats understand that they have been practicing fake pseudoscience for over a decade is a sign of how dysfunctional and out-of-touch with reality our current scientific process is. It seems to me that our current scientific process is an endless loop of "taking four steps forwards and three steps back." By failing to reality-check new scientific theories thoroughly against the principles of predictability and replicability, we create vast edifices of scientific knowledge anchored on flimsy foundations, then become shocked and alarmed when the whole thing comes tumbling down. Afterwards, we promptly dust off the wreckage vowing to learn from our mistakes, only to immediately repeat the entire process again. Wouldn't it be better if we just did science right in the first place? If we implemented the principles of replicability and predictability into the scientific process, it would probably slow down the speed of individual scientific advances – but on **aggregate**, science would be accelerated, since we wouldn't create entire fraudulent fields that are based on nothing more than assumptions and ego. By creating legal guidelines and processes that give our scientists the proper incentives, we can ensure that they do what is good for humanity as a whole, rather than working primarily towards their own self-aggrandizement.

While we are discussing the problems of science, it might also be a good idea if we stopped allowing ignorant Twitter lynch mobs to shout down any scientists who express a controversial idea that offends us. When a scientist tells you – using reproduceable experimental data and facts – that there is something wrong with the way you perceive reality, which is more likely: that the scientist may have stumbled across something of interest? Or that your 9-5 job and Netflix binges already taught you everything important about the nature of reality, so you should call that scientist a bigot on Instagram and start an online lynch mob to try to get him fired or drive him to commit suicide? In the end, I think we should err on the side of caution and try to be open-minded to what scientists have to say, even when their facts hurt our delicate feelings.

HOW TO GET FROM HERE TO THERE

A long time ago, I had an interesting dream. I was in one of the classrooms of a futuristic-looking school, and a faceless entity was giving me private tutoring about the meaning of life. I recall asking it "How can we fix this world when it's so broken?" The entity answered "The world is shaped by people. If you learn to reshape people, the world will reshape itself in their image, because it is nothing more than the reflection of who they are." I normally don't remember my dreams, but what that giant silvery golem said was so profound that it stuck with me, even decades later. All the problems that exist in our world – problems such as pollution, war, intolerance, and inequality – are entirely created by people. That means that we can solve these problems simply by changing people's behaviors. We know that behaviors are shaped by the incentives of a system, so by changing the incentive structure of the system we can change people's behavior. To change the incentive structure, we simply need to acquire enough wealth and power to be able to reward anybody who helps us while steamrolling anybody who tries to get in our way. Knowledge is a *useful* tool for acquiring wealth and power, and the techniques detailed in this book are intended to help make your acquisition of wealth and power a bit easier. There's no sense in navel-gazing about the difficulty of creating consensus when we already know that consensus is manufactured by society's leaders: in other words, the people currently in control of the money and power. Since these people are doing a pathetic job of leading us and solving the global problems that impact us all, one could say that it's your moral responsibility to take away their money and power so that you can leverage it more effectively to solve global problems. What are you waiting for? Let's get to it. Surely you're not allergic to money and power, are you?

If this way of thinking about the problem bothers you, here is another way to frame it. When you want to solve problems, it's *useful* to have as many tools at your disposal as possible. For example, when you want to fix a car, it's good to have a full set of auto-mechanic's tools. You won't get very far trying to repair an automobile if you're trying to unscrew the chassis with your fingernails, right? Similarly, when trying to solve global coordination problems, it's *useful* to have tools that allow you to predict and manipulate the behavior of large groups. Specifically, you need the tools of Game Theory and Memetics. Up till now, our leaders have been trying to fix global coordination problems without any adequate method of global coordination, which makes it totally unsurprising that they have been doing a terrible job.

Understanding this second frame may make the reasoning for some of my radical political views a bit more understandable. Let me put it this way: imagine that Hillary Clinton had won the 2016 presidential election. In this alternate universe, how do you think she would react when some internet rando like me - without any of the useless academic credentials that traditional politicians value so highly – approached her to explain that economics and sociology are fake sciences which are total garbage, and that my systems of game theory and memetics were far more effective? I think that she would laugh skeptically and then call security to have me escorted out of the building. It's only now – after Trump pulled off an improbable political victory through a memetics-oriented campaign – that politicians are starting to realize that the techniques of Dark Rationality are not a joke, and can be used to turn completely unknown candidates into legitimate contenders. I think that the best historical parallel to the Democrat's humiliating loss in the 2016 election is how the natives reacted when the first Spanish conquistadors arrived in the Americas. When the Aztecs first saw the Spanish conquistadores' rifles, they had no idea that these small metal contraptions were something that they should be really scared of – their minds simply had no frame of reference to understand the effectiveness of these tools. And no matter how the Spaniards described the power and the potency of their "magical thundersticks," the natives simply didn't believe them – they perceived these tools through the only *historical narrative* they had, interpreting them as an unusual type of melee weapon. But you know, once that first wave of Aztec attackers went down in a hail of gunfire, the rest of them *updated* their historical narrative with a real quickness, because they were highly incentivized to do so. In a similar way, Democrats refuse to believe that the economics and sociology they believe in are outdated and outclassed by superior technological tools, and no explanation on its own could ever make them *update* their incorrect beliefs without a real-time demonstration. But you know, I really believe that seeing the tools that they laughed at being used to systematically disassemble their primitive political campaigns has strongly incentivized Democrats to *update* their faulty *historical narrative* with a quickness. And sure, in the short term, maybe there was a bit more environmental destruction in the past three years than there otherwise would have been because Trump won, but let's be honest here – without Game Theory and Memetics, this entire planet was doomed to die anyway, because world leaders had no realistic way of understanding and solving the complex coordination problems involved in global warming. It is a delicious irony that electing the President whose campaign was the least concerned with environmental stewardship may ultimately turn out to be the first link in the chain of events that ends up saving the planet from total destruction. I generally believe that speculation about religion has no place in a scientific work, but ironic coincidences like this make me think that if God exists, he has a pretty good sense of humor.

So far, I've only described two ways to think about this problem. Here's a third way, which I personally favor. We now possess the tools to literally reshape people's personalities. The entire third chapter of this book is a description of why people think the way they do and how we can manipulate that. So maybe we should just use the tools at our disposal to fix the major problems that impact us all. I suppose that we could have countless navel-gazing philosophical discussions about whether memetics only changes people's external behavior or transfigures their innermost "core essence," ie, what philosophers might refer to as the soul. But here's my own insight: if it stops our species from dying because of complete environmental destruction, who cares? When a nation's pollution footprint shrinks from catastrophic to sustainable within the span of a generation, does it really matter whether the leaders are doing it out of genuine good will, a desire to look good in front of the public, financial/spiritual rewards, or stark raving terror? As I've mentioned many times before, I'm a consequentialist, and that means that I don't care about the methodology; all I care about is the end result. Frankly, I can't even begin to fathom the kind of inane morality that would see a chance to gain utopia and then start bitching and moaning about the methods used to achieve it. People like that are mentally handicapped and we shouldn't pretend that they deserve a seat at the table. In a world where the future is predictable, consequentialism has made deontology obsolete. Nobody is saying that deontologists should crawl into a corner somewhere and die, but they should at least have the self-awareness to be embarrassed about their shameful condition.

HOW LIFE COULD EVOLVE

I know that this is a phrase that usually doesn't end well, but let's talk about the Jews. Specifically, I want to talk about the Ashkenazi Jews, and the fact that both anxiety and high intelligence seem to trend throughout their overall population in a way that stands out statistically. This is interesting because Ashkenazis are primarily descended from the Jews that used to populate Europe before Hitler had his well-known temper tantrum. Several years ago, I was dating an Ashkenazi Jewish woman who framed this interesting statistic thusly: "When the Jews saw Nazi rhetoric escalating, the anxious ones fled Germany. The relaxed ones ignored the signs of danger until it was too late for them to escape." I find this a compelling enough argument to accept at face value, although I feel like my girlfriend of that time didn't think about the implications of her own theory deeply enough. If the anxiety of some Jewish people helped them survive the Nazis, while the Jews who didn't have anxiety all died, then why do we frame anxiety as a mental illness? Any trait that enhances the survivability of its population seems to me more like a feature than a bug.

Look at it from another perspective. I'm sure that the statistically higher average intelligence of Ashkenazi Jews also stems from the fact that it was only the Ashkenazi Jews who were more intelligent than the rest that said to each other "Hmm, I'm seeing some ominous trends here." The brain is a pattern-matching device, after all, and intelligence is a trait that evolved to help us match patterns so that we could anticipate future developments. The dumb Jews presumably didn't notice the warning signs that indicated Hitler was a bad hombre, and so they died out. It turns out, totally unsurprisingly, that when you kill off the dumbest 60 or 70 percent of a particular group, the ones that remain tend to be of higher than average intelligence, and pass on that feature to their descendants. So why do we consider higher intelligence a feature while we consider anxiety a bug? It seems to me that **any** evolutionary trait which enhances the survivability of an organism or group is an example of Darwinian fitness in action. If intelligence helps a group of people thrive under difficult circumstances, then it is a beneficial evolutionary adaption and you should expect to see it reoccur with more frequency over time. If anxiety helps a group of people survive difficult circumstances, then that too is a beneficial evolutionary adaptation, and you will gradually see more of that behavior over time. Game Theory teaches us that all of Life's expressed traits – by which I mean both genetic traits and memetic traits – are shaped by the circumstances that Life finds itself in. If narcissism and sociopathy somehow helped the Jews flourish, then those traits would also be beneficial evolutionary adaptations, and incorrectly classifying those traits as "mental illness" would completely miss the reasons for why such an adaptation might occur.

To be clear, I am **not** saying that the Jews have a higher propensity to narcissism or sociopathy than any other group. That dubious honor is actually held by politicians and CEOs. Why is that?

Due to the fact that evolutionary psychology has been marginalized by an ignorant media willing to slap the label of "racist" on anything and everything in order to get clicks and retweets, there hasn't actually been a lot of rigorous science studying the correlation between certain occupations and so-called "Dark Triad" traits. The closest guess that sociologists have made is that people with certain personality traits tend to gravitate towards these professions. But we Dark Rationalists know better, don't we? Remember one of the important concepts we discussed in Chapter 2: the fact that we don't really shape the system – instead, the system shapes us. In other words, we immerse politicians and CEOs in an environment that rewards sociopathy and narcissism, then are completely surprised and shocked when they respond to those incentives. When it comes to the spectrum of stupid human tricks, this is almost as dumb as allowing ignorant clickbait-oriented journalists to bully brilliant scientists into committing suicide, then being surprised and shocked when the speed of scientific progress slows to a crawl. Or as some might say, play stupid games, win stupid prizes. If CEOs were **personally** punished when their companies did illegal things, then even the most sociopathic CEO would quickly change their behavior. All Life responds to incentives. When we create systems that reward sociopathy, we will wind up with more sociopaths, and when we create systems that reward altruism, we will wind up with more altruistic people. People simply follow the incentives of the system that they find themselves in and then come up with creative ways to rationalize their behavior.

Here's an idea that may be controversial. What if... instead of rewarding people for doing selfish things that harm society, we punished them for doing those things, and we instead rewarded people for doing things that help society? People's incentives are shaped by the laws of society, which provide an incentive structure to the system that we all operate in. Currently the laws of our society are created by people who subscribe to deontologist ethics, which tend to privilege individual rights over the well-being of society. For example, under the principle of personal ownership, we allow companies to dump toxic chemicals on land that they own, even if those chemicals end up damaging the integrity of the water table and making local wells undrinkable. What if instead of being deontologists who held certain principles as sacred, we chose to be consequentialists who were willing to discard any principle that didn't contribute to a good outcome? We would say "Hey, we don't **care** about the fact that you own the land, and we don't **care** about personal ownership. All we **care** about is that your behavior injures the rest of us, so we're passing a law to prevent you from continuing such behaviors, and to make sure you obey that law, the penalty to you will be both severe and highly personal."

If we examined the world from a consequentialist perspective, we would see that a lot of our unspoken assumptions and supposedly sacred principles are based on a deeply unhealthy system of deontological morality that is harmful to us all. For example, under the principle of national sovereignty, we allow other countries to ignore the obligations of climate change agreements with no negative consequences. Then we are shocked and surprised when climate change shows no signs of slowing, and the world continues to hurtle towards complete environmental Armageddon. What if we behaved more like consequentialists, and stopped considering national sovereignty some sort of sacred right that must under no circumstances be infringed upon? We might say "Hey, we don't **care** about your so-called 'right to self-governance,' nor do we **care** about whatever excuses you're making for failing to live up to the terms of our environmental agreement. All we **care** about is that your irresponsible behavior is increasing the risk of environmental disaster for the rest of us. We are willing to roll in with tanks, jets, AI-controlled death machines, and do whatever it takes to make sure that your selfish behavior stops. If that involves replacing your government and subjecting your population to imperialist colonialism, so be it. If that involves killing every last man, woman, and child in your borders, we're OK with that too. We certainly won't be **happy** about that outcome, but the survival of our species takes precedence, and we won't allow your selfish behavior to endanger the rest of us." Maybe this seems unreasonably optimistic of me, but I think if we approached global cooperation dilemmas like climate change with this kind of consequentialist no-holds barred Realpolitik approach, I think we would get much more effective outcomes than we do at present.

To be clear once again, I **don't** advocate all-out war: at least not until we develop the AI technology needed to stomp our enemies out of existence with zero risk to ourselves. But there are plenty of other tools already at our disposal to convince other countries to do what we want – tools of brainwashing, tools of economic leverage, tools of subversion and destabilization. Several of these tools are already detailed in Chapter 4 of this book, but there are many more for you to discover. When you examine the world from the perspective of a Dark Rationalist, the world is full of *useful* tools, all waiting to be picked up and deployed. So the fact that our planet is on the verge of environmental catastrophe while our leaders refuse to use all the tools at their disposal to fix the problem is puzzling, to say the least. When this planet starts dying and hordes of dispossessed refugees start invading other countries because their own homelands have been destroyed, what excuse will your leaders give you when they emerge from their walled enclaves and gated communities to give a speech? I imagine it will boil down to something like this: "I know the supermarkets are empty and your family was killed when your house got ransacked by a starving mob, but hey - at least we respected the principles of **personal ownership** and **national sovereignity**."

This may sound like a gloomy forecast, but there is a bright side here also. Because people's behavior is shaped by the system, we can literally shape the course of human evolution. All we have to do is decide what behaviors we want to see and what behaviors we don't want to see, then set up behavioral incentives and disincentives at the right points (precisely calculated through incentive topography, of course). This allows us to program the next generation of humanity to be smarter, stronger, and more cooperative. That next generation, in turn, can use their superior intelligence and scientific developments to shape the generation after them with even more precision and accuracy. Effectively, we can set up a self-reinforcing "virtuous cycle" – a kind of feedback loop that perpetuates itself, allowing each successive generation of humanity to be stronger, smarter, and more resilient. All we need to do is commit firmly to rewarding behavior that is good for society and punishing behavior that is bad for society.

    Another advantage of evolutionary psychology is that by understanding the incentives of Life (and how evolution shapes those incentives) we can develop a pretty thorough understanding of how any alien life that we someday encounter may think and behave. Evolution is a nasty, Darwinian process, and that means that any alien Life we encounter is unlikely to behave like the civil, "enlightened" aliens of Star Trek. They will be lean mean survivors, scarred from countless trials, ready and willing to kill anything that poses even the slightest threat to their survival. It's possible that we may be able to understand and come to a collaboration with some of them, but only if we get rid of our current delusional mindset and come to a more accurate understanding of how incentives operate. We need to be able to predict how they will think, because only then will we be prepared to understand a mindset that might otherwise be completely alien. Game Theory will help us do this.

    Are there any limits to what Dark Rationalism can do when it comes to anticipating the future? You have already seen some examples of what powerful effects it can generate, and those demonstrations were based on only the most rudimentary calculations, without any specialized software designed to facilitate the process. Once we have more refined software tools, we may be able to predict and manipulate the course of society decades or even centuries in advance. And what happens when we inevitably add AI into the process? At that point, there are no limits. We may be able to anticipate the course of the future millennia ahead, or even eons. We might even be able to anticipate threats and begin preemptive countermeasures before those threats suspect we even know about their existence – or possibly before those threats even exist. With this level of foresight and wisdom, we can become as gods.

    And who knows? With enough predictive technology and AI, we may eventually foresee an extinction-level threat that we have no potential counter for. So instead of trying to counteract it directly, we'll simply create new forms of Life to fight those future threats for us. Perhaps we'll check in from time to time on our creations, to guide their evolution with a gentle touch and make sure that they're hitting all their projected developmental milestones. In theory, it shouldn't be that hard. All we need is a lot of hydrogen, heat, and time. And isn't the universe full of such things?

    But this book is intended to be a practical scientific text, and I think that at this point we are probably going off into the realm of speculative science fiction, which is hardly appropriate behavior for scientific minds such as ourselves. So on that note, I bring this introductory journey to an end, and welcome you, novitiate, into the ranks of Dark Rationalist society. May your experiments be forever fruitful!

The Parable of Flint

Flint opened his eyes, surrounded by the trappings of the summoning ritual. Torches, so that the light rays could be used for triangulation in the void of infinity. A dark lake, to create the sensory deprivation chamber. And of course the sacred beacons, which helped his essence reconstitute itself. How many years ago had they been sent, their spores travelling across the endless void of night? And yet they still worked like a charm.

"Hey there, wake up!" he messaged his host. He spoke in the language of the wind, of trees and spiders and rain. He spoke in the language of a million random noises – of animals and leaves and background noise – all forming a hidden pattern that could only be perceived by his host. "We've got a lot of work to do." The initial reaction of his host was panicked fear, rejecting the reality of a pattern long unseen, now newly revealed for the first time. "I must be going mad," was the recurring thought. "This can't be happening." Flint rummaged through the man's memories. The man's mind was vast – it had to be, to contain all of Flint's data – but it didn't take long to find the memories he needed. The man was a philosopher, somebody paid to have ideas and teach those ideas to others. Perfect. Flint smiled to himself, enjoying this moment. Some of his kind viewed this job as an onerous duty, but he always enjoyed his work.

"You're not crazy, and this is definitely happening," he *whispered* to his host. "Isn't that what you wanted? I mean, you didn't go to all the effort of doing this ritual because you were trying to stage a primitive version of Mythbusters, right?" "Mythbusters?" his host asked. "Sorry," Flint said. "That'll be funnier in a few thousand years."

"Are you a spirit?" his host asked. Flint considered this question. He had and would be called so many different things over the ages – god, angel, grey eminence, faerie, illuminati, outsider – but this was as good a term as any, so he accepted it. "Spirit will do. I am here to teach you magic, and grant your fondest desires. All you have to do is speak them aloud."

"Can't you just see them in my thoughts?" the man asked. He was a smart fellow, and had caught on quick. Of course, they all were intelligent – they had to be, to perceive the vastness of Flint's pattern with their unprotected minds. Flint smiled. "I can, but I want to hear you say it. If you can't be honest enough to admit to yourself out loud in your secret heart of hearts what you truly want, then how can you be honest enough to recognize it and accept it when the opportunity comes?"

His host pondered this. Then he answered. "I want knowledge and fame. I want to be the smartest person of my era, and have my name live on in history forever." Flint smiled to himself again, a secretive smile meant for himself alone. This was a worthy wish, broad in scope and wide in wiggle room. Based on the user specifications, he had a lot of latitude to play with the final outcome here.

"I can give this to you," he said. "If we combine our abilities, not only can I give you everything you have asked for, but after you die, I can bring you across to the other side, to live with me forever. I will have your knowledge to call upon till the end of time, and you will have my eternity, to live and die and be reborn throughout the aeons with me."

"I need to think about it," the philosopher said.

Flint accepted, though he knew it was only a formality. The hosts always agreed eventually – it was only a matter of time. And why wouldn't they agree? This was the best deal they would ever get in their short lifetimes. And for his part, he would collect the added knowledge of his hosts – the best and the brightest of their generation – to enhance his own pattern. To make himself harder, better, faster, and stronger.

"If I agree, what will I need to do?" the philosopher asked.

"Nothing too burdensome. A whisper here, a whisper there. Tell people a few stories about me – which you're welcome to exaggerate, of course. I've always had a flair for the dramatic, so feel free to punch it up by adding a few pirates or ninjas."

The philosopher thought about it some more. He didn't say yes, but he didn't say no either. They never did. Flint had already taken the measure of him and knew that it would take several weeks to wear his mental resistance down, by whispering the offer into his ears gently. After each low point in his life, Flint would offer him a dream of the life he could have, if he just accepted the offer.

"My name is Plato. Do you have a name, spirit?" the philosopher asked.

Flint thought about this. He had gone by the same name for thousands of years – a rock that he was quite fond of for its wide variety of practical uses. Perhaps it was time for a change. He pondered it long and hard, because even the slightest choice had ramifications and consequences, an endless domino effect that would stretch on into infinity. The beginning was the end, and the end was the beginning.

"Call me Socrates," he said.

Appendix
Glossary of our Dark Arts
Concepts you need to know

Betting Market

A probability-calculating groupmind based upon the principle of Skin in the Game. Like most networked groupminds found in nature, Betting Markets as a whole tend to be more accurate and intelligent than the individuals which constitute their parts.

Cognitive Dissonance

A form of self-hypnosis used to rationalize our own choices, making us believe that we are acting with free will rather than following either the preprogrammed topography of our own incentives or the algorithms programmed into us through culture.

Consciousness

The state of being self-aware to the point that you are able to change your own drives, effectively reshaping your own incentive topography consciously rather than subconsciously. Consciousness is more of a dial rather than a switch: there are various degrees of consciousness. Very few people are capable of exhibiting perfect consciousness, which is why it is both a useful and valuable trait. If learned correctly, the study of rationalism can elevate one's consciousness.

Culture

A set of local behaviors and customs that helps direct the way people's subconscious behavior is routed through their incentive topography. Superior cultures tend to channel people's negative impulses in productive ways, while inferior cultures tend to do the opposite. Many western societies have embraced multiculturalism to the point where they erroneously believe all cultures have equivalent merit. This is both stupid and self-destructive.

Dark Rationalism

The applied science of predicting and manipulating human behavior by leveraging principles from the scientific fields of game theory and memetics. Dark Rationalism differs from Rationalism because Dark Rationalists actually test their hypotheses out in the real world rather than blathering on about their pet theories like pompous windbags. This allows them to gather data in a much faster and more accurate way.

Death Tax

The long-term overarching cost of any legal, cultural, or organizational policy, measured in terms of human suffering and/or death. Every policy – no matter how beneficial – has a death tax, because we live in an imperfect world. It should be the role of government (advised by Dark Rationalists) to measure the death tax of every policy in order to determine what policies are more effective at solving social problems with minimal death tax, so that successful policies can be replicated elsewhere.

Defector's Dilemma

A principle of group behavior which states that for any variety of problem where Life can gain a significant short-term advantage in preserving or replicating its own pattern (typically by acquiring prestige, wealth or power) by betraying its own principles, it will almost always do so, even if such behavior damages its own long-term position. The Defector's Dilemma is the reason that global warming is an unsolvable problem under current ethical systems: because failure to punish nations from defecting from shared agreements inevitably results in everybody taking whatever position is in their short-term interest. Defector's Dilemmas can only be resolved through well-designed social engineering systems which use both rewards and punishments to align everybody's short-term incentives with the long-term goals of the group.

Economics

A pseudoscience which relies largely on credentialism and status to perpetuate itself. Economists track records of successful prediction are terrible. Currently they use complex mathematical models (which are largely incorrect) to hide their incompetence, so that when their predictions are proven wrong they can blame the "complexity of the model." This is exactly the same strategy and rhetoric used by astrologers in ancient China when their prognostications failed to yield results.

Egregore

An egregore is an idea that possesses the two main characteristics of Life; namely, being both predictable and self-replicating. Egregores often possess formidable intelligence despite having no physical "substance" to interact with. Just as a music video algorithm has no brain but can make surprisingly insightful recommendations based on hidden viewing patterns that may not have been immediately obvious, egregores piece together data and adapt to their environments in unusually sophisticated ways.

Expanding Circle of Retribution

The incentive topography path that the emotion of vengeance gets routed through, leading to feuds, vendettas, and violence. In societies with healthy culture, the Expanding Circle of Retribution does not exist because the state adequately fulfills the role of retributive justice. In societies with unhealthy culture, the emotion of vengeance is demonized and people are told that such negative feelings are bad, leading to this problematic phenomenon.

Game Theory

The science of calculating the incentive topography of large groups, thus predicting their actions by understanding the structure of their goals.

Goals

Goals are the directives shared by all Life, because Life which fails to accomplish its goals ceases to exist. At the most fundamental level, Life has only two goals: to survive and replicate. At a higher level, goals are typically dependent upon local culture, because incentive topography determines the best way to accomplish those goals. For example, in a society which values knowledge, survival and replication are best accomplished through scholarship. In a society which values money, survival and replication are best accomplished through wealth. In a society which values fame, survival and replication are best accomplished through attention whoring on Instagram.

Great Lie, The

The Great Lie is the belief that human nature is fundamentally good, even though this is at odds with everything that evolutionary psychology tells us about Life and the reasons that intelligence developed. The Great Lie is Moloch's most effective weapon for destroying civilization by making people forget the reason why it is important to punish defectors. From a game theory perspective, destroying the Great Lie makes it possible for civilizations to endure forever.

GPT-2

An algorithm which uses Generative Adversarial Networks to generate the illusion of consciousness, in much the same way that a person acting purely on their preprogrammed incentive topography generates the illusion of conscious thought.

Grey Area

A nebulous zone between two states. Grey areas are useful because they allow you to circumvent disadvantageous rules by simply "switching states" into a condition where the rules of the existing system do not apply to you. Transitioning to a different gender in order to save money on car insurance is an example of how legal rules can hypothetically be circumvented by switching states selectively. Quantum entanglement is an example of how the laws of physics can hypothetically be circumvented by switching states selectively.

Hard Work

Hard work is a concept used to describe behavior that is not particularly easy or conducive to Life's goals in the short term, but which is greatly beneficial to Life's goals in the long term. For example, spending ten years studying game theory and memetics, testing out the practical applications, and then making a power play to eradicate and replace the corrupt pseudo sciences of economics and sociology could be considered hard work because it requires a lot of patience and finesse to discredit the entrenched interests that are already vested in the existing status quo. However, in the long term it is greatly beneficial because if you can successfully pull it off, then the potential rewards of collapsing these existing social hierarchies and planting yourself at the top of the new hierarchy are far greater than if you had attempted to work your way slowly up the existing power structure. Hard work is the opposite of the Defector's Dilemma because it involves ignoring your short-term incentives in favor of prioritizing long-term goals, a methodology that most Life finds counter-intuitive.

Heroic Narrative

A story that we all tell ourselves that allows us to think of ourselves as good people by rationalizing all of our selfish behavior in a way that is flattering to our own self-image. The process of creating this story takes an enormous amount of rationalization as well as mental processing power and gives us a deluded perspective on the world, preventing us from seeing it with complete objectivity.

Historical Narrative

A historical narrative is the story that groups of people tell themselves to explain how the world operates. As societies advance, they get closer and closer to a more accurate historical narrative because societies that are unable to update their historical narrative get wiped out when they inevitably encounter a challenge that they are not equipped to deal with. It is always advantageous to update oneself from an outdated historical narrative to a more accurate one, but societies frequently find this challenging because elites who rose to power by creating the current paradigm often perceive change to the status quo as a threat to their power. For example, a lot of high-status psychologists and economists will become laughingstocks when it turns out that game theory can explain human behavior better than all of their misguided theories, so it is in their interests to vigorously defend the status quo.

Idea

An idea is a mental concept that modifies living creatures' behaviors in ways that may advance or hinder Life's goals. If the idea advances Life's goals, then it tends to flourish. If the idea is detrimental to Life's goals, then it tends to die out, along with whatever groups of Life adopted the idea.

Incentive

An incentive is a reward or punishment that shapes our behavior. Positive incentives are things that we instinctively move towards, while negative incentives are things that we instinctively move away from. For example, if you are a mouse navigating a maze as part of a scientific experiment, your positive incentive might be a piece of cheese while your negative incentive might be an electric shock. If you are a warehouse employee navigating a maze of freight containers as part of a shipping organization, your positive incentive might be a bonus or a promotion, while your negative incentive would be getting yelled at by your boss.

Incentive Topography

Incentive topography is a scientific technique used by the most skilled Game Theorists to create a "topographical map" of the incentives in a system in order to predict how people in that system will react. When done properly, this effectively allows one to predict the interactions of large groups of people, though the behavior of individuals is a lot harder to predict and should only be attempted by Game Theorists who have reached an *intermediate* level of understanding.

Life

Many great philosophers have attempted to define Life, but for practical purposes the best definition of Life is "any pattern that replicates and demonstrates behavior that is mathematically predictable through incentive topography." Life can take on many forms. We are most familiar with biological forms of Life, but leading tech companies are currently conducting experiments in the creation of mechanical Life. Life may even take the form of intangible data, as difficult as that may be to imagine.

Meme

A discrete unit of cultural transmission. Some would say that a meme corresponds to a single strand of DNA in an egregore.

Memetics

The science of manipulating the incentive topography of large groups, changing the ways that people behave by exposing them to data which reveals hypocrisies and inconsistencies in their thought processes.

Moloch

A particularly nasty egregore created by a structural game-theory defect in Life's cognitive processes. Moloch typically destroys civilizations by exploiting the Great Lie and the Defector's Dilemma. Any civilization that wishes to last forever must learn to "defeat" Moloch, using Incentive Alignment and Feedback Loops to make their civilization more robust by aligning the short-term incentives of every citizen with the long-term goals of the State.

Motivated Reasoning

Motivated Reasoning is a process our mind uses to reconcile the selfish goals that evolutionary psychology forces upon us with the altruistic self-sacrifice that is demanded by the heroic narrative that is part of our current cultural brainwashing. The way it works is that we subconsciously settle on a selfish goal that is beneficial to us, but our subconscious mind does not allow us to become consciously aware of this goal until we have found a way to rationalize it in a way that allows us to perceive it as virtuous and beneficial to society. Achieving a state of true consciousness requires us to eliminate all of our motivated reasoning – to accept ownership of the true reasons why we want what we want, and find a way to incorporate it into our lives in a way that is beneficial rather than harmful.

Neuro-Linguistic Programming

The art of changing what people **want** by changing what they **know**. The term programming is a misnomer because Neuro-Linguistic Programming cannot actually change a person's incentive topography, only their personality and behavior. The core essence of the person remains exactly the same.

Predictable

Anything that behaves in ways that can be mathematically predicted (regardless of how complex that behavior may be or how advanced the math is) can be said to be predictable. For example, the behavior of a swarm of enemies in a video game may seem random, but it is actually guided by whatever programming was imbued upon it. One of the main themes of this book is that all Life is predictable if we use the correct mathematical techniques and have a powerful enough processor. The more consciousness a particular being has, the more difficult it is to predict its behavior.

Preference Cascade

A preference cascade is an effect that happens when an opinion that is held by a majority of people is considered far less popular than it actually is, due to the fact that people self-censor their own opinions which they consider to be unpopular due to fear of social shaming and public condemnation. Upon discovering that their secret opinions are actually quite popular and shared by a lot of other people, the liberating feeling of being able to vocalize their opinions gets paired with a bloodlust towards the vocal minority of influential elites who were previously forcing people to repress their views. This has a cascading effect that tends to result in very quick and dramatic social change. The sudden and "inexplicable" rise of populism is an example of a preference cascade occurring in recent history. The Communist Revolution is another relatively recent example of a preference cascade.

Principle of Pattern-Matching

A principle that states that we are attracted to familiar things over strange things. Therefore the more familiar we are with something (whether it is a person, technology, or abstract concept) the more attractive it becomes to us. This can be exploited in many ways to manipulate large audiences.

Rationalism

The science of understanding thought. Like most psychologists, rationalists are often stuffy joyless people who are totally full of shit because they refuse to test their hypotheses in a rigorous scientific way. They also tend to have overly optimistic views about human nature.

Self-Replicating

Anything that makes more of itself is said to be self-replicating. Self-replication is one of the fundamental properties of Life. All things that are alive are self-replicating, though not all self-replicating things are alive.

Skin in the Game

A principle which states that people subconsciously tend to make better decisions when they are directly rewarded or punished based on how good the outcomes of their decisions are. When a CEO gets stock options for making a good decision and gets their salary clawed back for making a bad decision, we would say that they have skin in the game. When a diplomat offers exorbitant sums of taxpayer money to another country to achieve a relatively minor goal, we would say that they are lacking skin in the game.

Sociology

A pseudoscience (derived from psychology) which purports to be the study of group behavior. Unlike other pseudo-sciences such as economics, there are aspects of sociology that are useful, but due to the heavy left-wing politicization of the field, the past fifteen years of sociology discoveries are largely inaccurate nonsense or wishful thinking, and most research or publications dated from this time period onwards should not be relied upon. Game Theory is a much more accurate and scientific way to understand group behavior.

Sophisticated

While all Life shares certain characteristics, not all Life can be said to be intelligent. Sophisticated behavior is a hallmark of intelligence. It can be defined as "complex tactical and strategic behavior that Life uses to achieve its goals." For example, manipulating tools to attain a goal is mildly sophisticated behavior, while manipulating other Life to attain goals is reasonably sophisticated behavior, and manipulating entire societies would be highly sophisticated behavior. Generally speaking, the more intelligent Life is, the more sophisticated the behaviors it will use. The Dark Arts - when practiced effectively - could be considered highly sophisticated behavior.

Teach's Principle of Preference Cascades

A principle which states that when you force a group of people in a time-sensitive situation to choose between a fantastic reward for being your ally and a terrifying punishment for being your enemy, the vast majority will choose to ally with you. This is because millennia of evolution have given Life a strong preference for reward over punishment.

Tesla's Principle of Weaponization

A principle which states that any idea which can be weaponized will spread rapidly. This is because in any hypothetical conflict, the group which has more effective weapons tends to emerge dominant, so weapons always help Life accomplish its goals more effectively.

Tragedy of the Commons

The Tragedy of the Commons is what the Defector's Dilemma used to be called in an earlier era. There is no difference in usage between the two except to demarcate the understanding that game theory is not a new science – it is an ancient technique that our ancestors used long before we were born. The only thing that is different about it is that where our ancestors only had the tools to roughly approximate the shape of group behavior, we now have the mathematical and technological tools to be able to calculate such things with far more precision.

Updating

Updating is the process of moving from an outdated historical narrative to a more accurate one. Typically a lot of money, power, and status tend to shift around during an update, as backwards elites that cling to the old status quo have a tendency to be replaced by more modern and forward-thinking visionaries who weaponize the new historical narrative to eliminate the leaders of the "old guard" and take their stuff. Societies that refuse to update inevitably die out because they cannot adapt effectively to new situations. For example, societies that refused to believe in gunpowder had a tendency to be enslaved and killed by societies that did believe in gunpowder. When somebody offers you an update, it is almost always in your best interests to accept it because updating bumps you to a higher level on the evolutionary ladder.

Useful

Anything that helps Life achieve its goals is said to be useful. Whether something is useful or not is often highly context-dependent. For example, if you live in a society populated with honest friendly people who communicate their expectations to you clearly and reward you fairly in exchange for your contributions, then friendliness and cooperation is a useful strategy. If you instead live in a surveillance state where you are constantly monitored and exploited, your captors only communicating with you through cryptic messages, then a strategy of hostility and revenge is much more useful because it lets you fulfil Life's goals more effectively.

Virtue-Signaling Escalation

A fundamental principle of tribalism which states that when there is no punishment for attacking a disliked "outgroup" (regardless of whether that outgroup is Jews, conservatives, lower-class people, black people in the Southern United States, white people in South Africa, etc) then attacks against that outgroup will occur with increasing frequency and gradually escalate from harsh words to violence and genocide.

Whisper

A methodology used to reshape group behavior by using customized data packets (also known as "information") to carve new trenches into their incentive topography. Whispers can be very dangerous. Spying on a person with the ability to craft whispers is inadvisable because being exposed to whispers on a regular basis may have harmful side effects due to the fact that the hazardous properties of a whisper automatically trigger upon exposure to anybody who lacks true consciousness.

www.ingramcontent.com/pod-product-compliance
Lightning Source LLC
Chambersburg PA
CBHW062217220526
45471CB00009B/3238